Join the ATM App

C000203599

TO FIND OUT STUFF A̱
BANKERS DON'T KNOW…

- Who invented the ATM
- How it works
- Which secret technologies it uses
- Why ATMs are so vital to society
- How ATMs are crucial to disaster response and help combat global poverty
- Why serial numbers on banknotes are repeated
- How banknotes self-destruct
- Where to find the most extreme ATMs in the world
- And lots more …

It's fun! … it's FREE ! … and you can find it here:

theatmappreciationsociety.org

HOLE

in the

WALL

The Story of the Cash Machine

JAMES SHEPHERD-BARRON

"A fascinating personal insight into an important yet largely untold piece of social, financial and business history."

Hugh Carnegy, Former Executive Editor, Financial Times

In which
the son of the ATM's inventor,
tells the untold story
of how the cash machine was created;
explains the secret technologies
that make it work;
and explores its continuing contribution
to society
in a world where
physical and digital currencies
are colliding.

RUNWAY
PUBLISHING

First published in Great Britain by Kissyfish Books Ltd., 2017

Second edition, Published by Runway Publishing, June 2021

www.runwaypublishing.co.uk

Cover Design: James Lord

ISBN: 978-1-8384901-2-6

dedication

To my Dad,
John Shepherd-Barron
Inventor of the Automated Teller Machine.

My Dad's death, aged 84, in May 2010 prompted a letter from the then Chairman of Barclays – the first bank in the world to install a working cash dispensing machine – to his widow, Caroline, in which everything that needs to be said about both the ATM and my Dad is said:

> *"We all - by which I mean anyone who makes use of modern day banking facilities - owe him a huge debt of thanks.*
>
> *It would be difficult to overstate what a transformative impact his invention has had on the worlds of banking and commerce.*
>
> *It was a simply brilliant idea."*

John Varley
Chairman, Barclays Bank
May 2010

contents

prologue

FOR THE LAST TIME THAT SEARING Karachi afternoon, the honour-guard shouldered arms, turned to their left in one fluid military manoeuvre, and were dismissed. Rehearsals over, a dozen men in blue, sweat-stained uniforms moved gratefully to the shade of a hangar, propped their bayoneted rifles against the grey corrugated wall, and sank into the dry brown grass. As they fumbled for their cigarettes, the men cast furtive glances at a group of English nannies chattering happily under the branches of a nearby Tamarind tree, their small *Chota Sahib* charges playing noisily around their feet.

One of the men, shorter than the rest, with hair the colour of ripening corn and the brightest of blue eyes, detached himself and wandered over to introduce himself. The chatter slowed.

"Good afternoon ladies," the slight man said, removing his cap. His voice, languid yet measured, was unexpected. It was the voice of an officer, a gentleman, not the voice of an enlisted man in His Britannic Majesty's Royal Air Force. As one, the nannies rose to extend white-gloved hands, eyes peering curiously from under rims of felt cloche hats.

"My name is Shaw," he said, "Thomas Shaw, Aircraftsman, Grade 2. Pleased to make your acquaintance." One by one, it sank in. With a collective intake of breath they realised that the rumour must be true.

Blushing furiously, each nanny reached for the hem of her linen dress and began to curtsey.

Recognising that something unusual was going on, the small boys stopped playing with their toy soldiers in the dirt between the Tamarind's roots and squinted inquisitively at this unexpected intrusion into their carefree lives.

Embarrassed by the deference of the nannies, Shaw turned to ruffle the hair of the boy nearest him, asking at the same time if he would like to see a magic trick? Nodding enthusiastically, the boy gathered his playmates to sit in an expectant huddle at the feet of the grinning airman who quietly removed his tunic and began rolling up his shirt sleeves.

Pulling a small gold sovereign from his pocket, Shaw held the glinting coin up to the boys and explained with a wink how this very coin had been one of many used "in the Arab Revolt." Neither they nor the crimson-faced nannies had any idea what he was talking about. Bemused, they watched intently as Shaw proceeded to rub the coin vigorously against his bare, tanned forearm.

In a final snap of his fingers, it was gone, vanished into thin air. The nannies and the boys gasped. Unable to see how the trick was done, the boys clamoured for it to be repeated. Instead, they were told, "Behave boys! Say 'thank you', and shake the nice Mr. Shaw's hand ... and look him in the eye when doing so, as this is one handshake you will never forget."

To one small boy in particular, the one whose hair Shaw had ruffled, the memory of that day in the hot Indian sun with the sandy-haired aircraftsman was to be imprinted in his brain forever. Over eighty years later, he recalled this chance meeting with the nostalgic clarity of advancing age.

And with good reason.

Not only was this his first brush with the stuff that was to define his later life – cash, and all that went with producing, distributing and managing it – but he had just shaken hands with one of the most famous heroes of the 20th Century … *Lawrence of Arabia.*

The little boy was my Dad.

T.E Lawrence's exploits in the 1917 'Arab Revolt' against the Ottoman Empire – which was funded in part with 22,000 British gold sovereign coins carried by Lawrence – and his subsequent book, Seven Pillars of Wisdom, *made him an enduring hero the world over. In a futile attempt to escape the public notoriety which followed, he became a lowly aircraftsman in the Royal Air Force, enlisting under the assumed name of T.E Shaw. He wrote a book, published posthumously, about his time in the RAF, calling it* The Mint *to reflect how life in a training camp transforms raw recruits into functioning military men in just the same way that valuable coins are stamped out of metal blanks.*

The story recounted here many years later assumes a fateful irony when viewed in that historical light.

chapter one

atm

: ~'ôtə,mātəd/-'telər/-'məʃ,ën

noun

a semi-automated electro-mechanical banking device
which allows authenticated customers to repeat basic
financial transactions at any time without additional
external input from a bank teller, clerk or cashier

A GREY-HAIRED MAN WITH OWLISH spectacles and white
tuxedo tapped the unfamiliar microphone pinned to his lapel and
cleared his throat to speak. The murmur of 150 delegates slowly
subsided. The after-dinner speech was about to begin.

It's hard, looking back, to think this happened over ten years
ago, and that I was about to listen to my Dad speak in public for
the first – and last – time. It was his valedictory. He was to be
presented with an award that night for 'services to the ATM
industry' and we, his family, had been invited to share the
occasion.

Being family, we were seated on a VIP table towards the front.
On my right sat Dominic Hirsch, Managing Director of Retail
Banking Research, one of the most knowledgeable men in the
world on the subject of ATMs and the role they play in
commercial banking. On my left was an American who turned out
to be a director of one of the world's largest ATM manufacturers.

5

At one point, when my Dad was half way through his speech, this rather reticent and serious man suddenly leaped to his feet, pulled out his smartphone and started taking photographs. Within minutes, everyone else was at it, too.

Somewhat bemused by this unconventional behaviour, I turned to him once he had sat down to ask what this sudden passion for photography was all about? Guests don't normally go around randomly taking pictures of after-dinner speakers, do they? At least, they didn't in those pre-selfie days.

He didn't answer immediately, but then leaned towards me to whisper quietly but firmly into my ear, "Do you *know* who your father is?" His emphasis on the word 'know' confused me for a moment, but before I had a chance to reflect on what he was rhetorically asking me to consider, he went on: "I manage a division of a Fortune-500 company. All the people here this evening are the same as me. They are all executives in companies connected with money and how it is handled. These are not sales people celebrating their first commission." Still perplexed, and not being part of the financial services world, I waited for him to make his point.

"How often do you think we, or any business people, for that matter, get the chance to meet the man who created our industry, an industry worth billions of dollars and employing hundreds of thousands of people?"

I pondered for a moment the enormity of what I had just been told. In the background my Dad was winding up, seconds away from receiving the standing ovation he so richly deserved.

It was at that precise moment – one of those highly emotional, dumbstruck moments when words will not come out of your mouth – that I realised for the first time the immensity of my Dad's achievements … and the immensity of the man himself.

In a world where we have moved from cowrie shells and salt to banknotes, debit cards and mobile wallets in a single generation, the ATM story resonates in a way that could never have been imagined at the time of its invention. Belying its rather functional and minimalist exterior, it has become a legend in its own lifetime. This apparently simple machine has come further and evolved faster in fifty years than anyone involved in the early days of its development ever thought it would.

The ATM was 20 years ahead of the home computer.

Put together at a time when the moon landings were still two years away and the Beatles had just released their seminal hit single *Strawberry Fields*, the cash machine has gone on to survive the digital revolution and the threat of cyber-fraud to become as indispensable to our daily lives as the smartphone and the laptop.

It's only when we can't find one that we pause to consider what life must have been like before it came along. For all its cultural significance, few of us have ever taken time out from our hurrying, scurrying lives to reflect on how this lynchpin of today's retail payments ecosystem came about.

At the time, the idea seemed completely outlandish. Despite it being over twenty years since the end of the Second World War, the idea of automation was essentially the stuff of science-fiction back then. Bank managers, staff and customers alike thought the idea of the ATM ridiculous. "Why would anybody want access their money at 3 o'clock in the morning?" they would ask. "And why wouldn't they want to talk to a human being?"

But once the idea became reality, it quickly became clear to people that their lives were being transformed, and for the better. The shift was more fundamental than could be comprehended at the time and was to have far-reaching consequences for us as individuals, as well as for society at large.

According to David Lascelles, former banking editor at the Financial Times and now director of The Centre for The Study of Financial Innovation, "most of the high-tech, computerised machines we see around us nowadays owe at least something to the ATM." He makes an interesting point. It was, after all, the first machine with which ordinary consumers had to interact actively with technology in a direct, rather than indirect, manner; which involved users having to verify that they were who they said they were through the use of secret codes; and for which transaction records had to be kept. In this it was very different to, say, a chocolate bar dispenser or ticket machine." "It also," he reminds us, "pre-dated laptops, mobile phones and other conveniences we now say we can't live without by twenty years or so."

The ATM is much more than the sum of its constituent parts.

But does any of this matter? Is the ATM relevant anymore? Banknotes have been with us for hundreds, if not thousands of years, but do we still need them? Is the ATM, that icon of pioneering inventiveness and banking transformation, a manual typewriter in an era of touchscreen smartphones, an analogue dinosaur in a digital world? Worse than that, is the ATM an outmoded solution looking to solve yesterday's problem? Has its legacy actually turned it into an obstacle to innovation and progress?

The evidence would suggest not. Despite its lack of recognition, the ATM still plays too important a part in our lives to be considered a relic of last century's innovation. Unlike the telephone kiosk or dictaphone, it is far from obsolete. In fact, it can be argued that today's 'next-generation' ATM is even more relevant today than its fifty-year-old predecessor ever was, and to more of us than ever before. Not only can we access and recycle our money faster, more securely, and further afield, but its reach is extending to include the 2.6 billion people on the planet who have been denied access to financial services until now.

When the ATM came along, consumers experienced something meaningfully new ... not just an encounter with a new technology, but with an entirely new way of relating to that technology. The lightbulb did the same thing by converting night into day, thereby transforming not just our sleeping habits, but our work and leisure time. Laptops and mobile telephones have done it since, and virtual reality headsets are in the process of doing it now.

Yet, with the advent of new technologies like these, the novelty has us focusing on the way it works rather than what it's trying to do; we concentrate on the interface rather than the content until, with each iteration, it becomes more and more familiar, less and less awkward, and slowly it becomes an indispensable part of our daily lives. It was like this with the lightbulb. And it was like this with the ATM.

The ATM has always been more than a convenient way to get cash. Apart from the myriad other functions it can perform – including, with the more modern machines, the ability to cash cheques, conduct foreign exchange transactions, and recycle cash deposits – the ATM revolutionised the way banks think about their customers, the way the financial world manages risk, and the way we pay for almost everything we consume.

It's easy in the modern era of smartphones and tablet computers to forget just how strange it must have been fifty years ago for consumers to have such choice, and how novel it must have been to interact this way with a machine. People were used to asking for their hard-earned cash from a smartly dressed and

9

well-spoken human being behind a bank window barred with steel or made of reinforced glass. They wanted to see, with their very own eyes, each note and bill hand counted by a teller whose name they probably knew. They required a hand-stamped receipt indicating that their pay-cheque had indeed been deposited.

When the ATM arrived, it was met with great suspicion, and was considered by some to be more insult than progress, a ploy to reduce costs by getting customers to stop using expensive humans and to allow those staff who still had a job to do something more profitable. To the sceptics, it was as cynical as it was ground-breaking in its application.

Slowly but surely, though, the machine's blunt, no-nonsense exterior and intuitive user interface won people over, and the ATM began to blaze a trail in the way we interact dynamically with machines. After the technology earned the public's trust, an amazing change began to take place: face-to-face business became face-to-interface, placing consumers firmly on the path to a world of instant gratification and ever-more convenience.

The case can even be made that the ATM made e-commerce over the Internet possible. Some have suggested that it was our comfort with this innovation – an acceptance born of confidence that we were not being defrauded – which allowed the Internet and, later, online buying to gain such a hold on our imagination so quickly. Quite conceivably, Amazon and PayPal could not exist without consumers feeling comfortable interacting with a machine via a computer terminal ... which is exactly what an ATM is. "The ATM is the mother of all this," says Bob Tramontano of the US consumer technology company, National Cash Register. "It was the ATM that established this trust between user and machine."

The world of financial services also owes a debt to the outsourcing of day-to-day banking to ATMs by proving to banks that there were alternative distribution channels to bricks-and-mortar branches, and by opening the door to online and mobile banking. In this sense, it was the ATM that enabled the explosive growth of retail finance during the last decades of the 20th century.

But it was doing more than that at the time; it was revolutionising the way we think ... mostly because it allowed us to buy things without having to plan ahead. It also banished the age of deference and put us in charge. Quietly, insidiously, and without our really being aware of it, the ATM's reliability ensured it became a trusted friend, and in so doing it became a touchstone of the millennial era.

The mid-1960's was a time of ordinary people emerging from a post-war world of rationing and austerity; of a struggle for hearts and minds of the common man and woman who, possibly for the first time, had the chance to wonder about the possibilities of what might be. It was a time of challenge and change. An ideological time where paradigm shifting touched every aspect of people's lives almost every day. The advent of the ATM did not just serve as a milestone on that journey of discovery, or even mark that day when money became our own again, but, through the mainstreaming of a semi-automated self-service culture, it allowed us for the first time to determine for ourselves how our precious time could be managed.

I don't think Paul Volcker, former chairman of the US Federal Reserve[1], had that in mind when he once famously opined that "the ATM has been the only useful innovation in banking for the past 20 years." Since he uttered those words, two things have altered the landscape of retail payments: The first is that our trust

---

[1] The Federal Reserve is not a US government entity. It is a privately owned central bank owned by the banks that are members of the Federal Reserve system. It is no more a part of the US Federal Government than Federal Express.

in banks and those who run them has been steadily eroding. The second is that technology has allowed our mobile phones to offer a digital alternative to the physical currency we call cash[2].

Despite these shifts – perhaps because of them – the ATM of today remains one of the most trusted points of contact in a rapidly changing financial ecosystem. Early ATMs were brutal metal monsters, intimidating to look at and clunky, impersonal, and inflexible to operate. They were also stand-alone affairs, unconnected to anything other than the electrical supply of the bank that housed them. Initially, they could perform only one, albeit primary, function of a human bank cashier or teller … dispense cash. They were activated by plastic or paper tokens which were only distributed after suitable creditworthiness checks – usually involving an interview with the bank manager – and which were then returned to the customer by post once the account had been debited. These tokens would only work in designated machines owned and operated by the customer's bank, too.

Since then, ATMs have become much more sophisticated and nowadays can perform almost all the functions of a human teller. What's more, they can do this at all hours of day and night, every day of the year, and in some of the most remote and inhospitable places on the planet. And, unless they have run out of the folding stuff, they do it mostly without fail. It is this reliability that drives our trust in the ATM.

This reliability was no accident. Over-engineering the ATM so that it would work without fail was one of the three founding principles driving the ATM's early development (the others being its ability to prove that the person receiving the money was who

---

[2] Currency is the word used to describe a system of money in general use in a particular country in the exchange of goods or services for cash. Today, the term generally refers to printed or minted money.

(s)he said (s)he was, and that an accountable audit trail could be maintained).

Nevertheless, despite the constraints of limited functionality, it's surprising that it took more than a decade for banks to deploy ATMs at scale. This was probably because, in its early days, few believed that the ATM would make much of a difference to the average consumer. As its obvious convenience steadily changed patterns of consumption, however, the number of ATMs grew, enabling unplanned impulse buying at weekends and at night. At the same time, it allowed retail banks to grow their customer base by giving them access to customers who'd previously been excluded. The nature of work in bank branches also began to change, as employees were shifted away from cashier services and into sales.

For a long time, ATMs were also expensive to install and difficult to operate. The need for dedicated phone lines still limited them largely to bank branches or high-traffic locations such as train stations and airports. Hardware and software systems had to be custom-made which didn't make them any less costly. As with all other aspects of retail financial technology, this only changed with the advent of digitisation and the introduction of common software platforms as the core operating system. It was only when these 'back end' components came together that the ATM effectively became a terminal of a bank's central computer, thereby enabling almost instantaneous online authentication at the point of use. Updating central records from the point of a transaction is easy in today's world of mobile banking and e-commerce, but the ATM was the first device to be truly networked into more than one bank's central server, and in real time too.

Whether the 'cash dispenser' deserved to be called an 'automated teller machine' has been questioned by some in light of these early limitations. However, given that it was certainly automated and that dispensing cash was the most important function of a bank teller back then, most industry observers now consider the acronym to be an accurate, even if rather limited, description of what the machine then did.

Whichever way you look at it, the ATM and its inventor deserve an enormous amount of credit – or blame, depending on your point of view – for enabling us to live the way we do today. With so many people over the years laying claim to being the machine's progenitor, though, it's hard to know to whom the credit should go. Perhaps, like knowing who invented tin-foil, the matter of a machine's invention is of little consequence? The ATM, after all, "just is". Perhaps knowing who had the vision to put it together is no longer relevant. But, you could say that about the lightbulb. It just 'is' too. But in the days of gas, oil, wicks and soot, it wasn't. It took someone to flick a switch and bring us out of the dark ages.

If you ask anybody which 20th century invention has had most impact on our daily lives today, most people would probably suggest the mobile phone or the personal computer. It is doubtful that people would think first of a machine nestling innocuously in our urban landscape which dispenses cash. Yet since it emerged in the mid-60's, the ATM has become – and forgive the pun – part of the currency of our lives. With the pressing of a few buttons, money can be transmitted and conjured in seconds from almost any ATM anywhere in the world. The ATM made accessing our money faster and more convenient, and we could do it further and further afield. Even in the early days, it is difficult to conceive how revolutionary this was.

Like all great inventions, the idea of 'fast money' didn't take long to ripple around the world, and it was this that marked the real beginning of the cash revolution going on around us today.

In fact, so far has the ATM come in fifty years that the acronym no longer accurately describes what it does, as it operates more like a bank than a bank teller. An ATM may be just a cash dispensing machine in most people's minds, but it's obvious the device has a lot more to offer, even in the digital online age.

chapter two

pioneer
: ~' /pīə'nir

noun
one who is first or among the earliest to begin or help
develop new areas of thought, research, or
development in any field of enquiry or enterprise

A COUPLE OF YEARS AFTER DAD DIED, my brothers and I were rummaging through his belongings trying to decide what to keep and what to throw away when we came across a large, bulging envelope. It was one of those old, yellowed colonial-era envelopes where the flap is tied down with string. Untying it, an enormous pile of silvered sepia photographs tumbled out onto the floor. They were of a time gone by; of a dusty Karachi landscape, immaculate lawns, spit, polish, and nervous servants standing at attention. Picture after tiny curling picture showed empty beaches, clear skies, long shadows, unfettered privilege ... and immense opportunity.

None of the three of us had seen any of these photographs before. One series was particularly noticeable. They show a boy, perhaps three years of age, on a donkey peering uncomfortably at the camera from under the rim of an over-sized 'pith' helmet. Dressed in jodhpurs and riding boots, the poor boy is clearly sweltering in the hot Indian sun, his sweaty little round face radiating nervousness. He has no interest in the donkey except for the fact that he is sitting on something that could explode into bucking revolt at any minute.

For the boy, it is triumph enough that he has ridden the donkey without disgracing himself by falling off. His look says it all … he will be relieved when it is all over. It is a face desperate to please an over-achieving and upwardly mobile mother, but fully knowing that, whatever he does, it will never be enough. Something of this dynamic can be glimpsed in the photograph at the back of the book of Dad on the same donkey taken nearly three years later. At the donkey's head stands his mother – my grandmother – impeccable in Edwardian hunting attire, smilingly beautiful but at the same time aloof and imperious.

Striving to live up to impossible expectation is what the British and the stiff-upper-lip was all about in those days. Dad no doubt suffered from all this, and certainly said of his parachuting days in the Army some eighteen years later that he deliberately chose the most hazardous of military pursuits to convince his mother that he was no nerdy wimp. This sort of insecurity is not uncommon in lonely children, and often influences their later ambition to succeed.

He was an unlikely hero. Born on June 23rd, 1925 in the ebbing days of the British *Raj*[3] in Shillong, India (now Bangladesh) Dad was surrounded by hard work, hard play, ambition and inventiveness from an early age. His father, Wilfred Barron[4], a rather dour Scots engineer from Wick, built some of the largest commercial sea-ports in the world including those of Aberdeen, Karachi and Chittagong, and helped design the World War II

---

[3] The term *Raj* refers to British rule in the Indian subcontinent between 1858 and 1947 and extended over almost all present-day India, Pakistan, and Bangladesh.
[4] Wilfred Barron changed his name to Shepherd-Barron on marrying Dorothy Cunliffe Shepherd, and later became Chairman of the Port of London Authority and President of the UK's Institution of Civil Engineers.

Mulberry Harbours[5] for the D-Day landings in Normandy. Each of these feats of engineering employed innovative techniques on a grand scale.

His mother, Dorothy Cunliffe Shepherd – or 'Dolly' as she was affectionately known on the international tennis circuit – was a Wimbledon tennis champion. Although she died in a car crash before I was born – tragically with my grandfather at the wheel – she was obviously one of those gregarious and socially ambitious socialites who was, as her son put it himself, "too damned good at everything." As the wife of an Indian Army Officer, for example, she had no sooner picked up a rifle for the first time, than she immediately became 'All India Female Rifle Champion.' Even more impressive is that she only took up tennis a few months before reaching her first quarter finals at Wimbledon.

She got better at tennis though, and went on to win a bronze medal in the women's doubles at the 1924 Olympics and became the ladies' doubles champion at Wimbledon in 1931 … much of this helped by practicing in Monte Carlo with the then UK Minister for Defence, Sam Hoare who would think nothing of flying her there for the week-end from London at the taxpayer's expense in an RAF transport aircraft. She also played in Norfolk, England with the King of Sweden, and in Los Angeles with the likes of Charlie Chaplin[6].

How she came to marry a dour Scots military engineer was something of a mystery to Dad all his life, since they only seemed to share one mutual love which was horses. But, there again, Dad didn't really know his mother very well. Until the age of thirteen

---

[5] Mulberry Harbours were developed by the British during World War II to facilitate the rapid offloading of cargo onto beaches during the Allied invasion of Normandy in June 1944. They are still considered one of the greatest military engineering achievements of all time.
[6] Charlie Chaplin was the first person ever to install tennis-court lighting. His first game under lights was played with Dolly as his partner against Douglas Fairbanks and Hazel Wightman.

he lived in India, returning to the UK every year by P&O liner with his nanny, 'Nanny Danes', to spend the Summer with his maternal grandmother in Norfolk while his mother gallivanted around the European and American tennis circuits. From the age of eight, he, like all small boys of privilege in those days, was packed off to boarding school.

Their paths did overlap in Norfolk, however, in the summer of 1932, when she let Dad, then aged seven, play a few strokes with her mixed doubles partner, 'Bunny' Austin, the most famous British tennis player of his era … as famous then as Roger Federer is now, at least to the Brits. After seeing that the long cricket flannels 'Mr. Austin' was wearing were soaked with sweat and clearly slowing him down, Nanny Danes recalled the precocious John asking him after the knock-about, "why don't you wear shorts like me, Mr. Bunny?"

Perhaps it was just coincidence, but Austin travelled to Forest Hills the very next day with Dolly on the trans-Atlantic ocean liner *Queen Mary* where he scandalised American society by wearing shorts for the very first time. The next year, he continued his practical fashion statement and broke all protocol by wearing shorts at Wimbledon too. If it was more than coincidence, then this was Dad's first, albeit indirect, brush with innovation.

A few years later, he was in the kitchen with Nanny Danes at home in Karachi when a horde of white-dhoti-wearing locals came to the back door demanding to borrow every pot and pan in the house. With the exception of one pan – to boil the *Chotha Sahib's*[7] milk – they were handed over. Dad recalled that the man politely doing the asking wore round wire-framed spectacles. Later that day, every one of the pots and pans was returned.

Having shaken Lawrence of Arabia's hand a couple of years earlier, Dad had just met another of that era's iconic figures,

---

[7] Hindi for 'little boss'.

Mahatma Ghandi, who was at that time leading a peaceful 'salt march' revolt which was to culminate eventually in the departure of the British from India[8]. This, coincidentally, was also his second brush with money – the stuff that was to define his later life – as salt was considered currency at the time (see the later chapter on *Cash* for more on this).

Invention is the product of a curious mind at work.

Dad was educated from the age of eight in English boarding schools. From his Preparatory School at Rottingdean in Sussex, he moved, aged thirteen, to Stowe School in Buckinghamshire, where, by his own account, he "idled away" his five years as World War II came and went playing tennis, some of it with his older friend, C.R.

C.R was son of the author A.A Milne and was none other than the Christopher Robin immortalised in the famous Winnie-the-Pooh books. Already a celebrity in 'his' own right, Dad remembered Pooh-Bear[9] being taken everywhere by C.R … "even to lessons. He was once put in the umpire's chair to referee a crucial inter-house tennis match". This very same stuffed toy bear can now be seen – rather the worse for wear – in the New York City library.

On leaving Stowe, Dad was commissioned into The Royal Artillery – the 'Gunners' – a few months before the end of the Second World War. He was originally posted to an Air Landing Regiment in Germany, but was diverted a few days before embarkation to accompany 100 enlisted soldiers to India. After

---

[8] In March 1930, Mahatma Gandhi deliberately violated the British salt law by scooping up a saucepan of muddy seawater and proceeding to manufacture his own. He then implored thousands of his followers to do likewise, which, over the months ahead, they duly did. With this simple act, the fate of the British Empire in India was sealed.
[9] Christopher Robin Milne's original toy bear, Winnie-the-Pooh, is now on display at the main branch of the New York Public Library in New York City.

three weeks at sea by way of Southampton and Bombay, he was immediately posted to Bilaspur in the Central Provinces as a Troop Commander in 555 Parachute Light Battery, part the embryonic 159th Parachute Light Artillery Regiment, itself part of the 2nd Indian Airborne Division.

Parachute instruction was given at Chaklala airfield outside Rawalpindi, now in Pakistan. Here, Dad and his 'Ghurka Gunners' spent a happy two weeks jumping off towers and out of airplanes. In one incident, a nervous Ghurka was pushed out of the aircraft taking half of Dad's trousers with him. On landing, the embarrassed Ghurka presented his instructor with a sacred kukri[10] by way of atonement.

His pet Dachshund *Popski*, the regimental mascot, came into his own at Chaklala. When jumping out of airplanes, it helps for the pilot to know which way the wind is blowing, as otherwise the deploying troops may drift off course to land kilometres downwind. To gauge this, a weight suspended under a small parachute would be thrown out of the aircraft on the first pass over the intended drop-zone, the drift of which would be noted by the pilot, allowing him to make the necessary adjustments to his second 'run-in'. Using a modified cargo parachute and harness designed by my father and the senior packer at the airbase – the second of my father's putative 'inventions' – Popski became the 'regimental drifter'. This involved him being thrown out of the plane attached to a static line which would automatically deploy the parachute, to drift slowly, barking all the way, to earth. Being a Dachshund – the famous 'sausage dog' with a long tummy and short legs – his tummy would hit the ground before his legs, at which point he would be released from the harness by the

---

[10] The kukri – or khukuri in Nepali – is a Nepalese knife with an inwardly curved blade. It is used as both a tool and as a weapon in Nepal and is the cherished weapon of the Nepalese Army.

automatic quick release button deliberately designed to hit the ground first, thereby freeing him to run back to where he could by then see his master landing. This was sometimes some way away. "Never once did he get lost," Dad once recalled, " ... and every time, a local would eventually appear at the main gate to hand back the parachute and harness." Popski's quick-release mechanism, adapted from what the Ghurka parachutists were already using, was Dad's second brush with invention.

No sooner had parachute training been completed, than Japan surrendered. While his troops rejoined the rest of the regiment, Dad was sent to Burma where he spent three weeks accepting the surrender of small numbers of Japanese troops guarding what by then were empty prisoner-of-war camps. By the time his short posting was over, he reckons to have collected more than twenty Japanese Samurai swords, which, with no-one wanting to take responsibility for their disposal, lay in a growing pile under his bed. He used to joke that, at one point, he must have had the largest collection of ancient Samurai swords in the world. Unfortunately, history doesn't relate what was to befall them, as Dad was hurriedly posted back to Karachi where his regiment was now moving, pending deployment to Quetta in Baluchistan, and had to leave them behind.

The British were concerned that the Russians would take advantage of the relative chaos at the War's ending by invading India through the Khyber and Bolan passes on the western border with Afghanistan, as Genhgis Khan once had. Accordingly, a parachute Brigade was being hurriedly put together, and it was to Brigade Headquarters that he had to report. Eventually, he found where they were ... located in his old family home at Clifton, just outside Karachi.

Before embarking for Quetta, one incident in particular – which has since entered the annals of British military infamy – kept him busy.

This incident involved the quelling of a mutiny in the Royal Indian Navy. This relatively new navy had gone through a period of rapid growth during the war, and now consisted of 28,000 officers and men, most of whom had seen wartime service. Its ships were on the elderly side and confined mostly to escort duties[11]. On 9th February 1946, one of these ships, a corvette, was moored in Karachi port – alongside the very quayside built by my grandfather – and took part in the mutiny which started with HMIS Hindustan further to the south in Bombay and had quickly spread to Calcutta and beyond. It must be remembered, not only that the War was over, but independence was looming. The British Raj – and, with it, the British Empire – was nearing its end, and everyone, Dad included, knew it. Initially, the British thought it best to treat "the disturbances" as a strike, but with HMIS Hindustan opening fire on Indian soldiers and training its guns on the Yacht Club, a bastion of British prestige, across the harbour, a different view quickly took over. Dad takes up the story:

"Locking their officers below, the ship pooped off her 4-inch guns and sank a tug, plus landing some random rounds in the centre of Karachi. Eventually, the Commander-in-Chief in Delhi, a General called Claude Auchinleck who had been replaced by Montgomery in the North African desert campaign of '43 – and with whom I later had a fascinating evening of reminiscences in Morocco in 1974, which I regretted not recording for posterity as the old boy died shortly afterwards – got tough, and ordered two companies of Ghurka parachutists and a couple of our 3-inch guns into action. We

[11] Trevor Royle: The Last Days of the Raj; Penguin, 1989

24

had permission to fire if surrender had not been forthcoming by a certain time.

About half-an-hour before the agreed deadline, the mutineers manned their 4-inch gun, and trained it in the direction of one of our 3-inch pack howitzers, the barrel of which was poking out from behind the corner of a *go-down* (shed) 125 yards away. As gunnery officer in charge, I had laid the gun with cross-hairs – i.e in the direct fire mode – and knew we could not miss.

They had three shots, all of which passed harmlessly over our heads. There didn't seem to be any explosions, probably because the fuses had not armed themselves at that short range. It also seemed to me that the gun could not depress any further than it had, and that we were below their firing line and therefore out of harm's way. I shouted this to my men, but they weren't convinced. Before the fourth round could prove me wrong, we were given the order over the radio to open fire. We disabled the gun and killed the crew with three well-aimed rounds, and then, because nobody told us to stop, pumped one round of phosphorous into the wheelhouse, which set the ship ablaze. They surrendered pretty quickly after that.

This was such a rare event in British military history – the army and the navy having a go at each other – that I wrote to the Commander Royal Artillery requesting our gun section be given a medal. My commanding officer was furious, complaining that the incident was too embarrassing for officialdom. I lost the battle. But, curiously enough, was promoted to Captain shortly afterwards."[12]

With the unsatisfactory trials that followed the ending of the naval mutiny, and the failure of a UK diplomatic mission shortly

---

[12] Handwritten fax by John S-B to James S-B in October 2005

afterwards, it was becoming obvious to many that India was quickly becoming ungovernable. It was against this backdrop that Dad made his way to Quetta.

No sooner had he settled in that Dad persuaded his Battery Commander, Tom Carew – a former Special Forces officer who had seen service behind enemy lines in both Europe and Burma, and who was therefore not adverse to a little 'adventure' – to add skiing to his men's fighting skills, arguing that the local terrain might call for action above the snow line. With the mountains of the Karakoram range looming behind him, it was difficult for Tom to say, 'no.' Having duly got his permission, Dad set off with five brother officers for three weeks of ski training at Gulmarg in Kashmir in sight of the majestic Nanga Parbat.

On his return, a new form of parachuting was about to be trialled, and Tom, recognising Dad's have-a-go attitude, selected him to run it. Given that the terrain they were supposed to be guarding against Russian invasion – not to mention marauding Afridi and Baluch tribal activities across the border with Afghanistan – was either completely flat or full of jagged mountains, some 'bright spark' had come up with the idea that parachute troops could be lobbed over mountain ranges. The technique would require the launching aircraft to gather airspeed by diving to the valley floor of one valley then pulling up sharply, thereby imparting momentum to the parachutists in the back who would leave the aircraft at the appropriate moment as it was in the climb. This, so went the theory, would enable the parachutists to "freefall upwards" – this was the oxymoronic term used – over the mountain tops, and begin their decent down the other side into the next valley, at which point they would pull their ripcords. As if this idea was not hair-brained enough, the hapless parachutist's helmets would be loaded with lead in order to impart even more momentum.

Dad thought the idea completely mad, and told Tom so. The plan went ahead, however, and Dad and six Ghurka parachutists duly took off early one autumn morning to put it into effect. What he hadn't told anybody observing the drop zone on the ground, though, was that each man had secretly made a dummy of himself, to which, as the aircraft lifted off into the brightening sky, they now fitted their parachutes, each of which had been secretly fitted with a pressure-sensitive automatic opening device. Dad recollects that their fear at the time was that they would either pile into the mountainside at speed on the way up, or have too little height on the other side for safe deployment of their 'chutes. If the dummies made it, he and his comrades would then try it out.

As the red light turned to green, the men fixed the weighted helmets to the dummy's heads and shoved them off the ramp at the back. Dad remembers this being harder than it sounds, as the g-forces and buffeting slipstream were considerable. He then went forward to sit with the pilot to hear over the radio what the result had been on the ground. The pilot was expecting to make multiple run-ins, so was not surprised by this. He was, however, somewhat more surprised to notice that all six parachutists were still on board. Just as he was making his next run-in, the radio crackled into life and ordered the aircraft back to base. The trial was cancelled.

It was not until the de-brief some time later that they found out what had happened. Dad, a war-time officer who was about to leave the Army and return to complete his university studies, remained thoroughly unconcerned by what his 'failure to obey a direct order' might mean, as his promotion prospects were of little consequence to him. Soon enough, a chastened Tom Carew walked into the room carrying one of the dummies … or what was left of it. All had gone perfectly. The dummies, matching the weight of their human counterparts, had sailed serenely over the mountain ridge; the parachutes had deployed perfectly; even more amazingly, each of the six had landed squarely on the intended drop-zone. This was, in itself, remarkable, as Dad used to joke that the troops were quite used to being dropped into entirely the wrong valley by the RAF.

The only problem was that all the dummies had landed without their heads. It transpired that the helmets, which had been weighted with quantities of lead, had snapped back as the parachutes deployed, and the neck had not been able to withstand the whiplash forces that resulted. Since the dummies were those used for bayonet practice and therefore made of metal frames, this must have been some force. The idea was not tried again. Dad received a formal reprimand from his commanding officer, "but noticed a mighty wink as I saluted and turned on my heels to leave his office." Later that evening, he was once again formally presented with a kukri from the five grateful Ghurka's whose lives he had unwittingly spared.

Not long after this happy event, the day of India's independence was upon them. After centuries of misadventure in the sub-Continent, the British were finally leaving. "One dramatic day," Dad wrote years later in response to having seen Tom Carew's obituary[13], "we paraded our 24 guns, complete with jeeps and limbers, and manned by their British gunners, on ceremonial parade in Quetta. The drama of the moment was not lost on any of us. At precisely midday, the Ghurka pipes and drums played our Regimental slow march, and the Union Jack on its pole in the centre of the parade ground was slowly lowered. The Colonel gave the twenty-four six-man crews the order, 'Six paces, March!' leaving room for the same number of Indian Army gunners to march into the places vacated by us. They then mounted, and

---

[13] Letter from John S-B to the Secretary, Perse School, Cambridge, England, 27 October 2009

drove the guns around the parade ground in troop sections of four. It was all quite emotional, considering the long history of the British Army in India.

Amid the political tensions, Dad prepared British troops to hand over their equipment and used to joke that he had, "pulled down more Union Jacks than Mountbatten." He also joked – not without irony – that he was the "last son of the Raj", and in many ways he probably was.

Within days the Regiment sailed to join the British 6th Airborne Division in Palestine. For the British, Palestine was a salutary lesson. The Army had been pushed into a campaign ill-prepared for an insurgency, and military wins in 1947 had failed to translate into wider success when it became clear that the government had neither the will nor the authority to continue governing[14].

With each move, the violence between Arab and Jew escalated, as did the number of incidents against British targets. The execution of nine Jewish members of the Stern Gang, convicted of terrorism by British courts, was carried out in July 1947. Reprisals against the British quickly followed, including the kidnapping and murder of two British sergeants. This, in turn, sparked a rampage by British troops in Tel Aviv – causing five more Jewish deaths – and widespread anti-Semitic violence and rioting in the UK.

Into this upheaval, landed Dad, then, so he thought, on his way home to demobilise from the Army[15] and complete his university degree, this time at Trinity College, Cambridge.

Unfortunately for him, his arrival in Palestine coincided with that of 4,515 illegal Jewish immigrants aboard the now infamous

---

[14] Jonathon Riley: The Babe; published privately, 2006
[15] John S-B was one of the last officer intake to be drafted under wartime provisions, and was never a Regular Officer.

ship *SS Exodus* which had been escorted into Haifa by the Royal Navy the day before. No sooner had they arrived, than London, in another U-turn of British policy, decided that the would-be settlers were to be returned on British vessels to France, from whence they had sailed. Dad was chosen as the Officer-in-Charge of British troops on one of these vessels. After refusing to disembark in Marseilles, they were taken on to Hamburg where they were off-loaded by force and returned to camps in the British-occupied zone of Germany.

This sorry saga was played out in full view of the world's media, and also happened to coincide with a UN mission sent to come up with recommendations on the future of a Jewish State. Some historians think that the way the British mishandled the whole affair may have contributed to the UN's recommendation on partition, and subsequent US support for this. If so, Dad had become another small footnote to history.

The irony was not lost on him. Here he was, having been born in what would become Bangladesh; having been present at the birth of a newly independent India (later to be partitioned into Pakistan); and to have been involved in the eventual birth of another nation only a matter of a few weeks later. "This," he once said, "truly was the end of Empire."

Security Express: Europe's first
armoured trucking company.

After some more to'ing and fro'ing in this long passage from India, Dad eventually made his way to Cambridge, 'coming down' in 1950 with a fairly modest degree in Law and Economics. Shortly afterwards, he joined De La Rue, the world's largest security printing firm, as one of their first-ever management trainees. According to his brother, Richard[16], he was lucky to get the job[17]: "As there many applicants, the company decided to hold some competitive interviews, and all the candidates were told to get an early night so they would be nice and bright on the day. John, however, went to a party, overslept the next morning, and arrived late. Lucky for him, the procedure was over-running, and, as he was fortuitously nearly last on the list to be interviewed, his late arrival passed unnoticed. The then Chairman decided he would take one candidate to lunch, and picked my brother."

Dad also thought he was lucky, but for another reason: A man named Alan Houseman, tipped to be the next Chairman of De La Rue, was on the interview panel. During the course of the interview, it became clear that Alan's father, once a doctor in Shillong, was probably the doctor who had delivered him!

He spent the whole of his working life from that moment on with this one company. This was not unusual at the time but is difficult for those competing for jobs now to comprehend. "Such was the stuff of loyalty in those days," as Dad used to say.

Trained initially in Thomas De La Rue, the original currency division, Dad's first appointment was as export sales manager for Formica[18], mainly covering Australia and Africa, and where he spent most of his time negotiating the setting up of dealerships and agencies. Despite the extensive travel involved, he somehow had time to meet, and marry my Mum, Caroline Murray. This can't have been easy, as she hailed from a family who lived in the far north of Scotland. But, whatever the logistical challenges, he had chosen someone with a lively mind and for whom

---

[16] Richard S-B, born twelve years after John S-B, is a world-renowned racing driver, having won his class in the 1953 Le Mans 24-hour race in a Morgan; the only person ever to do so in his own car.
[17] From Richard S-B's eulogy made at John S-B's funeral in Tain, Scotland, May 2010
[18] Formica® is a kind of plastic laminate countertop that became popular in the 1950's and 60's.

inventiveness also ran in the family: A succession of grand-uncles had, between them, designed and constructed a new type of plough – the commemoration of which can still be seen today carved into the side of a monument near Lairg in Sutherland –; a new type of combination valve which is still used to blend the correct mixture of oxygen and acetylene for the cutting of steel – an invention which some say gave rise to the second industrial revolution –; and the world's first salmon ladder, a series of water-filled steps which allow migrating salmon to ascend rivers otherwise blocked by hydro-electric dams.

After two years, he was promoted to run the Stationer's Division. For the most part, this involved manufacturing playing cards. In fact, De La Rue was known best by the public for this aspect of its business. Years later, I could never understand why our house was always littered with cellophane-wrapped packs of crisp new playing cards, and, as a family, we did seem to play a disproportionate amount of card games compared to our friends … I suppose for no other reason than the cards were there. Probably because this job didn't interest him very much, Dad pretty quickly recommended that this division be sold to Waddington's, the maker of board games, which it duly was.

In 1957, with me not yet two months old, and with instructions telegrammed to his long-suffering wife instructing her to sell everything they owned and follow him, he was sent to the US to assess opportunities for the company in the North America's. On arriving in New York, one of his first commercial ideas was to make the discount vouchers distributed with Persil washing powder to look like real money, an idea which saw a leap in sales as soon as it was introduced. To do this required the establishment of a subsidiary corporation which he called The De La Rue Banknote Company. This name, combined with his early success with Persil put him in a position to negotiate a contract for

De La Rue to print stock certificates for the New York Stock Exchange, the first time an overseas company had been allowed to do so. They were printed under high security conditions in the UK and flown to New York.

Towards the end of his two years in New York, he evolved the idea of organic growth within the company in a way that covered all aspects of 'life of currency' and recommended moving into every area of currency handling, from printing banknotes to handling and processing, distribution and destruction. Nowadays this would be referred to as an 'end-to-end' approach but it was a completely new way of doing business back then. He went further, and proposed to the parent company, the Thomas De La Rue Currency Division, that "We not only move cash around in armoured vehicles but build the vaults at either end of the supply chain and even install the alarm systems." In short, he was recommending the sort of 'vertical integration' that would require a security printing firm to transform itself into a full-service cash management company.

This approach was accepted in principle by Sir Arthur 'Gerry' Norman, the then Chairman of De La Rue. So, having seen the potential for armoured trucking – a concept then unknown in Europe – and having secured the backing of the Board, Dad set about finding out what he could about the way Brinks Matt, then the largest 'cash-in-transit' company in the US, operated. This research ended up with him proposing a joint venture with Wells Fargo, eventually resulting in the two companies investing a modest £80,000 between them. At this point, Dad returned to the UK to set up Europe's first 'armoured car' or 'cash-in-transit' company which he called *Security Express*.

The company was 51% owned by De La Rue and 49% by Wells Fargo (whose parent company was American Express) and was set up partly to cajole the latter into giving De La Rue the UK contract to print their Travelers' Cheques[19]. The first of two trucks was on the road in February 1960, but growth was slow until a monopoly was achieved to carry all the London clearing bank money on a daily basis throughout England and Wales. The big Scottish clearing banks were more reticent, however; a situation which quickly changed after the Great Train Robbery[20] of August 1963, when, after an overnight visit to Edinburgh, Dad secured the contract to move their money too. The truck fleet then grew, "as fast as the Commer Vans could be converted", and eventually numbered 633 in all. Amex was later bought out for the princely sum of £81,000; an investment that turned into £16 million by the time the company was sold to Australian firm Mayne Nicholas fifteen years later in 1982. Dad was rightfully proud of this achievement.

As Chairman of Security Express, he still spent time regularly in the US, and on one of his visits had taken a long hard look at the fledgling Federal Express (FedEx) operation started by Fred Smith from a hangar in Memphis, Tennessee. Dad spent a few days with Fred and after a few days working with him asked what he thought of De La Rue starting a similar, but road-based system in the UK. Fred's reply was that, "as long as it doesn't compete with me here in the US, I have no problem with that."

Blatantly copying the idea, an overnight parcel delivery service called *Courier Express* was consequently established in 1970 as an offshoot of Security Express and based around the existing depots. It was a success from the start, owing to a single contract with the British clearing banks. The idea involved picking

---

[19] A Traveller's Cheque is a medium of exchange that can be used in place of hard currency when abroad. Their use has been in decline since the 1990's as alternatives such as debit cards and ATMs became more widely available.
[20] The UK's Great Train Robbery took place in the early hours of 8 August 1963 when a 15-strong gang, including the notorious criminal Ronnie Biggs, robbed a Royal Mail train heading from London to Glasgow and got away with £2.6 million (equivalent to over £55 million/$70 million in today's money). The money was never recovered.

up cheques cashed that day from every branch of every bank in England south of Manchester, taking them to London to put them through the central IBM cheque sorter, and returning the output tape to each and every branch before opening for business at 9:30 a.m the next day. This business model cut the clearing time for cheques from three days to less than eighteen hours. Despite the extra income generated by the banks, who could now put the money saved onto the overnight money markets for two days in a row, the cost savings were never passed on to the customer. This rankled with Dad, who, as with the ATM, saw the model as being a win-win for both the banks and their customers, as transaction costs were dramatically reduced in both cases.

Dad remained chairman of Security Express but was appointed as Chief Executive (CEO) of De La Rue Instruments in 1964. The company had just one product, a currency counter developed in Sweden, with one Japanese bank the only customer. Even that contract was to be short-lived, as the Japanese quickly discovered the machine lacked patent protection. So they did what the Japanese do best, copied it, improved it, and made it themselves. This was done with honour, though, as, despite being under no legal obligation to do so, they paid a license fee to De La Rue throughout the transition period.

The only other revenue stream came from being European distributor for a Russian-designed high-speed currency sheet counter used for production control purposes in the printing of money.

All in all, a very limited market, and attempts to add additional income streams to the rapidly dwindling business of five full-time and six part-time employees were becoming rather urgent. On his own recommendation, Dad was given two years to turn the company around.

The most likely initiative was the UK's first cash-voucher acceptor which De La Rue had just fitted to 400 Shell garage forecourts to speed up the delivery of petrol. Sadly, though, when the trial was over, the idea was rejected on the grounds that it took the public twice as long to fill their cars – remember, these were the days of petrol pump attendants, who doffed their caps and cleaned your windscreen while you remained in your vehicle – and the Fire Services were not happy with the risk involved. The voucher acceptor worked perfectly, however.

More successfully in terms of generating a sustainable income stream, Dad persuaded the Royal Mint to buy a vacuum-coin-wrapping device[21], most of the profits from which were derived from the amount of paper and cellophane the Mint wasted in the production process and which it had to buy exclusively from De La Rue.

But, while revenue improved over the two years that it took to bring these products to market, the balance sheet remained precarious, and it looked increasingly likely that Dad would have to wind the company up. A new product was needed, and needed fast.

---

[21] This deal involved becoming an agent of Molins Corporation.

chapter three

invention

: ~' in'ven(t)ʃ-(ə)n

noun
a new, useful and not obvious machine or process that
did not exist previously and that is recognised as the
product of some unique intuition or genius

LOOKING BACK ON IT NOW, it must have been the Easter
holidays. Two things make me fairly certain of this: One is that
the day, being bright and clear, was bitterly cold. The other is that
I was on my way to see the dentist in London. As with most small
boys I imagine, this annual event was not looked forward to
much. But the dreadful anticipation was always tempered by the
promise of some compensatory special event. Usually, this would
be a West-End matinée, a visit to a museum, or, best of all, a
couple of hours across the Thames at the Battersea Fun Fair. But
this time, it was different. For a start, there was only me. And
second, my Dad was taking me to the Royal Mint "where money
is made."

I only found out about this during our fifteen minute drive
into Cobham station that morning. Mum had dropped us off in
the station wagon – a gigantic gas-guzzling legacy from our early
years in the United States – with a parting rejoinder to my Dad
not to let me eat anything until after the ordeal with the dentist

was over. Dad, of course, promptly ignored this advice, buying me a chocolate bar from a vending machine on the platform as we waited in the hustle of the commuting melée for the London-bound train.

Later, as the train clattered and rocked its way through the dreary landscape of commuter-belt Surrey, I asked Dad why we were going to the Royal Mint [22] and not Battersea Fun Fair. Secretly, I was rather disappointed. Watching money being printed didn't sound like much fun. Folding away his newspaper, he leaned forward to explain, in an almost conspiratorial whisper, that I was "in for the treat of all treats ... that I was to see how ten shilling ('ten bob') notes [23] were made ... and would be the first boy ever, in the whole of England, the United Kingdom, and the whole world, to do so."

Satisfied with this, I sat back and unwrapped my chocolate bar. I must have been more or less the same age as he had been when photographed with his mother riding a donkey in India.

The tooth-ache was forgotten within minutes of arriving at The Mint which, in those days, was housed in a grand Georgian building near the Tower of London. First, I watched an old man wearing eyeglasses – I recall his name being Fred, but I might be wrong – finish engraving the die which was to be part of the printing process. While letting me hold the die – making me potentially the richest boy in the world, or so I thought in awe – he explained how 'ten bob' banknotes required multiple such dies, all created by hand by master engravers who never meet each other so as not to be able to divulge any of the secrets of their work. Then, under the watchful eyes of uniformed security

---

[22] The Royal Mint has an unbroken history of British coin production that dates back over 1,100 years to the reign of Alfred the Great.
[23] Before decimalisation, one Pound Sterling comprised twenty Shillings, with each shilling made up of twelve Pence. Shillings were known colloquially as 'Bob', a term used since the 1700's.

guards, I witnessed a 'top secret' test print-run[24]. Fred explained the vagaries of 'ink and rag (paper)' while this was going on, but much of what I was being told went in one ear and out the other while I focused, fascinated, on what one of his colleagues was telling me about the 'security devices' that would eventually end up being incorporated into the final product.

What I remember more clearly than any of this, though, was Fred's explanation of why all banknotes in the world not only have unique serial numbers, but why these numbers appear twice on the same note ... (an historical anomaly which is explained in more detail in the chapter on *Cash*).

Returning home late, my Dad and I went shopping the very next day, a Saturday, and it was then that the second, seemingly innocuous, event that would change our world forever took place.

I had accompanied Dad to the bank in the hope that an advance on my pocket-money might be forthcoming because I was itching to buy the 'must-have' fashion item of the time, a 'superball'. It wasn't the closed doors that made me realise my hopes were about to be dashed, but the fact that this was the first time I ever heard my Dad swear. Since I didn't know what the 'F-word' meant at that time, it wasn't the word itself that shocked me so much, but the vehemence with which he said it. I remember him grabbing my hand and marching me up the Cobham high street in a tower of rage to a garage whose kind owner cashed him a personal cheque instead.

These two, seemingly disconnected, events played an enormous part in defining the ATM's story.

---

[24] This 'test' was on a stand-alone intaglio press, and was done by hand, one note at a time. Bulk printing of the actual banknotes destined for circulation was done at the time of my visit in the mid-1960's at De La Rue's plant at Debden in Essex.

Invention is more perspiration than inspiration.

As with most aspects of clarifying 'who invented what, and when', the answer to the question "who invented the ATM?" is not as straightforward as it first appears and is not without controversy. There have been arguments for years over who should officially go down in history as the ATM's inventor. The answer depends in large part on how the question is interpreted. It depends, in other words, on how the words 'invent', 'automated' and 'teller' are defined. It also depends on the timelines of history, the prevailing cultural attitudes of the day, and your views on whether it takes more than one person to invent something.

Before examining this evidence – some of it not seen before – it's first worth considering what constitutes an 'invention'? Although referring to another age, this historical anecdote provides some context which should help us work out a suitable definition for ourselves:

With the possible exception of paper money, the plough, or the wheel, few inventions have influenced human history more than the stirrup. Historians believe that the competitive advantage afforded by this simple piece of technology – a hoop of metal attached to a horse's saddle with an adjustable leather strap – gave rise to a new form of mounted warfare against which foot soldiers were no match. This most basic of technologies conferred such advantage to mobile mounted cavalry that traditional forms of set-piece warfare were rendered obsolete almost overnight.

It doesn't take much imagination to realise that early bareback riders, especially those festooned with weapons of one sort or another found mounting a horse something of a challenge. Without stirrups and with only the most rudimentary of saddles, they spent a lot of energy trying to get on ... and, once mounted, trying not to fall off.

40

Then, around three thousand years ago, someone added a simple loop of leather to the saddle that allowed the less athletic and more heavily armed rider to mount his horse more easily. From there it was simply a matter of time before someone added the loop to the other side and a new advantage was found ... the rider could not only mount more easily but was now much more stable into the bargain, and therefore less likely to fall off in the heat of battle. Before this simple invention, only highly skilled and well-practiced horsemen were able to ride and wield weapons effectively at the same time.

But this wasn't all. By 700 AD, Europe began to develop a new form of society whose sole purpose was to support this highly mobile type of mounted warfare, eventually structuring their entire society around the concept of the mounted knight. The material cost of maintaining mounted knights was staggering and the feudal system evolved as a way of supporting this new type of warfare.

No-one is suggesting that the invention of the ATM is on a par with the stirrup, but there are some parallels: First, it was new. Nothing like it had existed before; second, a simple and elegant technical innovation provided an efficient and cost-effective solution to a problem no-one knew they had; third, entire societies adapted to the new technology; and fourth, the technology itself evolved, giving rise to yet further consequences which had not been foreseen originally.

With this in mind, we have to ask ourselves what exactly it was that Dad was supposed to have invented?

Any meaningful discussion of this subject would necessarily have to start with defining what is being discussed. For purists, a cash dispensing machine and an automated teller machine are different. One definition suggests that a cash machine is a stand-alone device requiring human intervention whereas the ATM is effectively a remote terminal of a bank's computer network[25]. This would mean that the ATM as such could only have been invented after the invention of linked computer networks ... the first example of which, the Ethernet, only went public in 1977, ten years after the cash machine was launched. This would be akin to saying that the light bulb could only have been invented once the electrical supply system had been put in place. But it wasn't; it was invented decades earlier. If we insist on making the distinction, however, it took about two decades for the cash machine to transform into the ATM.

The ATM certainly met the criteria of a 'machine' as it was an externally powered mechanical apparatus made up of several different parts, each with its own definite task, yet which performed a particular function when working together, and did so over and over again. It was, to an extent, also 'automated' in as much that, once the user's identity had been authenticated, it dispensed cash without further human input. But in its original guise it was unconnected, even to the bank in whose wall it was installed, and was therefore in a sense, dumb. Nor did it make use of the magnetic-stripe card still in use today ... for the simple reason that magnetic stripes hadn't been invented by then.

And whether or not it performed the functions of a bank teller – otherwise known as a 'cashier' or 'clerk' – depends on what you think a bank teller actually did back in the mid-sixties. Officially, their job was relatively straightforward: To pay out and take in cash over the counter. In practice, they spent most of their time counting the stuff. Well, the ATM performed that function really well: it dispensed cash. Input of a personalised token, when matched with a four-figure personal identification number in real time, resulted in cash being delivered to an authentic customer

---

[25] Batiz-Lazo: Was the ATM a disruptive innovation? ATM Marketplace, 21 January 2016 [accessed October 2016]

whenever – and, to an extent, wherever – he or she wanted it. Furthermore, it was delivered correctly and quickly, safely and securely, day and night, come rain or shine. No bank teller could do that, and no precursor – including CitiBank's Bankograph which had debuted in New York a few years earlier, or the Bank of Japan prototype trialed in Tokyo in 1966 – had been able to get anywhere close to achieving such a successful outcome before.

Invention and innovation are not the same.

If the identity of the person who first came up with *the idea* of a working cash dispensing machine is not in dispute, the identity of the person – if indeed there is such an individual – who transformed the concept into something more akin to the multi-functional connected ATM we recognise today, can legitimately be debated.

Modern invention, as we will go on to see, is a blur of competition and collaboration. Deserving individuals – such as John Glenny of De La Rue who did so much to pioneer application of the personal identification number, or PIN – are too often left out of the fray, becoming casualties in a form of pseudo-intellectual triage driven more by ego than historical fact. The patent record doesn't always help.

The man who holds the patent to the device in the US is Don Wetzel, credited in the Smithsonian National Museum of American History in Washington DC as the machine's progenitor. In 1995 the Smithsonian recognised Docutel and Wetzel as the inventors of the ATM[26]. Wetzel, however, did not hit pay dirt on his 'invention'. His name was on the patent (along with two others), but Docutel owned it. "I never got any royalties," he says. "But I was treated very well by my company."

---

[26] Diebold and Fujitsu were developing ATMs at this time, but Docutel was the first to patent.

The same cannot be said of De La Rue and Dad. Despite later successes with Security Express and Courier Express, he was never rewarded in any way by this very conservative and risk-averse British company, and although everyone thinks he must have been fabulously wealthy on the royalty payments generated by his invention, he never received more than his salary and pension. As a 'salary man', he never thought it should be any other way.

The man who holds the patent for security aspects of the device in the UK is James Goodfellow. But the ATM's roots can be traced further back to someone named Luther Simjian, who put together a machine that allowed customers to deposit cheques and cash. He persuaded New York's First National City Bank (now Citibank) to give it a try back in the early 1960s, but customers had little use for it, mostly because it wasn't very reliable. "It seems the only people using the machines were a small number of prostitutes and gamblers who didn't want to deal with tellers face to face," he later wrote.

As recently as October 2016, James Goodfellow was inducted into the Scottish Engineering Hall of Fame as "Inventor of the ATM". A few months earlier, in April 2016, the Guardian newspaper in the UK ran an article also citing him as the "unequivocal inventor of the ATM". This would appear to be a logical conclusion. He was, after all, the patent holder. But, as with everything else in the murky world of 'inventology', the truth is somewhat different.

In the Guardian article, Goodfellow talks about his role, including how, as an employee of an engineering firm in Glasgow, he was tasked with designing and patenting a machine which would issue cash on demand to a recognised customer at any time of day or night, seven days a week. The solution involved coupling something only the user knows – the PIN –

with something the user owns – an associated coded token – sent to them through the mail by their bank branch. This token took the form of a plastic card with holes punched in it. When inserted, a card reader matched the position of the holes to the sequence in which four of ten buttons mounted on the machine's facia were pressed. "This pretty much describes the ATM today," the article correctly points out.

After little more than eighteen months of testing, the first Chubb-branded machines to use Goodfellow's system were installed at branches of National Westminster Bank (later to become NatWest) in August 1967. This was nearly two months after the De La Rue version was unveiled.

So, what was it about the original De La Rue machine that made it first, and made it so different? As with the Chubb machine, customers could withdraw ten pre-packed one-pound notes using a special token, though the De La Rue version took the form of a paper voucher. These vouchers were deliberately designed to look like cheques as customers were already familiar with this method of getting at their cash. They were supplied in packs of 10 and issued free to approved customers only, each of whom was sent a personal identification number (PIN) in the mail. Six digits were used during the testing phase, but, on the advice of my Mum, who said six was too many for her to remember, and my eight-year-old brother who said that four would probably be enough anyway if a 'three-strikes-and-you're-out' rule was applied[27], this was reduced to four by the time of the ATM's launch.

---

[27] This was my older brother, Nicholas, who was always something of a maths prodigy. Having spent his formative years in the US, he was also familiar with Baseball's 'three strikes' rule. He is now a professor of mathematics at University College London and a Fellow of Trinity College Cambridge and The Royal Society.

A single voucher was placed in the drawer of the ATM – or "robot cashier" as they were known by bank staff at the time – which the customer then closed. It took about 20 seconds for the machine to verify the validity of the voucher after which a green light on the facia blinked on. Seeing this, the customer then used the keypad to enter his or her PIN which, if it matched the coding sequence on the previously inserted voucher, illuminated another green light to signal that the drawer had been unlocked and that the packet of cash which had replaced the voucher could now be extracted.

The authentication technology came in two parts: The first was kept extremely 'hush-hush' for obvious reasons and involved the use of mildly radioactive ink when printing the voucher. Once inserted, a small Geiger-Muller counter verified that the voucher was genuine. The press were told that the ink was not radioactive, but magnetic[28]. One reason for this minor subterfuge was a fear that the public would not accept the use of radioactive ink when memories of the nuclear bombings of Hiroshima and Nagasaki were so fresh. The other reason was to deliberately lay a false trail for anyone intent on misappropriating the technology. This was especially important as the whole concept was not under patent protection at this time. But we will come onto that.

The second – and parallel – method of authentication was more visible and involved a sequence of holes drilled through the voucher through which a light was shone. Encoded within the sequence was the four-figure PIN which could be 'interpreted' by a simple array of photo-receptors. If the numbers were correct and, crucially, entered in the correct sequence, the drawer release

---

[28] As early as October 1965 – or almost two years before the ATM's first deployment – both the Times and Financial Times in London reported that banks were in the process of developing "automatic machines (that will be) giving cash at any time", and described a system that operated with magnetic ink vouchers.

46

mechanism was activated once more, and the customer could open it to find the voucher replaced by a packet of cash.

In essence, the only difference between the two machines, then, was that the De La Rue machine didn't use a plastic card, preferring instead to apply technology it already knew about through its involvement with automated petrol pumps … perforated paper cheques overprinted with mildly radioactive Carbon-14 ink[29]. In the article, James Goodfellow accepts he didn't invent the concept of a cash-dispensing machine. Nor did he invent the accompanying four-digit personal identification protocol we know as the PIN, as the whole concept had already been passed to the Kelvin Hughes engineers by Tom MacMillan of Nat West[30]. What he did invent, however – and go on to patent – was a new, more sophisticated, and more elegant way of applying it using Hollerith 'decade' switches[31]. In this, he was like James Dyson … he had invented a better mousetrap.

The same article went on to suggest his involvement with the ATM was similar to the Wright Brothers in that they didn't invent the concept of flying either … everyone was trying to do it. But they achieved the first successful powered flight and so earned the credit for inventing the aeroplane. To add insult to injury – and as contributors to the newspaper's blog later pointed out – this whole affair was akin to calling a train driver the inventor of public transport. He was employed by a company whose client, Chubb, had contracted his firm to come up with a way of

---

[29] Carbon-14 is a soft-Beta emitter and is therefore mildly radioactive. It has to be ingested in enormous quantities to cause any harm to human health.
John S-B hand-written note drafted for the BBC (date unknown)
[31] Herman Hollerith was the creator of the Hollerith Electric Tabulating System, the ancestor to computers as we know them today. The 19th century system used cards with punched holes to tabulate data. As a primary form of data input for computers, the punch card existed into the early 1970s, well before the interactive display terminal began its ascendance. Hollerith died, aged 69, in 1929 in Washington, D.C., and is widely known today as the father of information processing.

automating the dispensing of cash to legitimate recipients, so he did.

Sadly, although he was no doubt unaware of this at the time, he was applying secret ideas that had already been developed by someone else ... the Barclays-De La Rue team. It is therefore a mistake to claim, as the Guardian newspaper has done and as the Home Office in the UK is also now doing in its 'nationality test', that James Goodfellow is the ATM's inventor.

The ATM's development was a team effort.

Interestingly, Dad himself never claimed to be the ATM's 'inventor'; as has already been mentioned, that accolade was pinned on him later in life after a TV documentary by the Discovery Channel in 1995 referred to him in those terms. This was then further substantiated in The Nilsen Report, the Guinness Book of Inventions, and the ATM industry association's own newsletter. The theme was repeated by various TV and radio channels thereafter, including by the BBC in 1997.

But could the ATM ever have been the brainchild of a single person anyway? The argument against the 'great man theory' of invention is not new and suggests that it wasn't. In his book *The Innovators*, Walter Isaacson points out that most innovation is not the product of one person acting alone. He cites the case of the man credited with having conceived the first automatic digital computer, Charles Babbage, and his collaboration with Ada Lovelace, the world's first computer programmer, without whom his Difference Engine would not have been so very different. His collaboration with Gottfried Leibnitz, the most famous German mathematician of his day, and Alan Turing of Bletchley Park and Enigma fame is highlighted, as is the fact that the British government supported his work to the tune of £170,000, then equivalent to twice the cost of a battleship. Someone somewhere had seen merit in the idea and had 'sponsored' it. The book also explores what Ada Lovelace called "political science" ... that machines and their creators are not just about cold calculation and mechanics, but about the essence of human creativity. "Those who helped lead the technology revolution were people in the tradition of Ada, who could combine science and the humanities," he writes. Innovation is all about getting the mix right.

Dad himself was always the first to say the ATM's development was a team effort. However, as a businessman – not an engineer in the traditional sense like his father – it is true that he came up with the idea, proved the concept, found a 'sponsor', and led the team at De La Rue that re-engineered existing components into a new and innovative electro-mechanical device. As such, he wanted to be remembered as a business pioneer rather than inventor. And, just for the record, he actually thought his pioneering work in setting up Security Express, Europe's first armoured trucking service, and setting up the UK's first overnight parcel delivery service, Courier Express, ranked higher in his list of achievements. He certainly thought they were more fun.

How is a business pioneer defined? Is an inventor necessarily a business pioneer? Should Alexander Fleming be regarded as a business pioneer because his discovery of penicillin led to the mass production of antibiotics? Should Thomas Edison, a prolific inventor whose company grew into General Electric, be considered one? Clearly, just inventing something isn't enough. The invention – or innovation – has to have real-life application and should have upended old ways of operating and led to the creation of a new business.

There is inevitably a backward-looking component to deciding who is a genuine business pioneer, and who is not. By definition, a pioneer breaks the mould by doing something different, the true importance of which might be hard to identify until years, or even decades, later. Early iterations of the ATM involved refining the process of dispensing pre-counted packets of cash, and it was fully 15 years before its wider potential began to be realised. This retrospective component also tends to favour western industrial economies, not least because of their ability as free-market societies to communicate each successful innovation through advertising and the media. Western economies also have strong legal and patent systems to protect and nurture transformative ideas. The Wright brothers were just as determined to protect their flying machine from unscrupulous competition through the courts as they were in learning to fly. The same was true of Edison's lightbulbs.

Meanwhile, historians now like to point out that Marconi did not invent the radio any more than Edison invented the lightbulb, or Charles Darrow invent the most popular board-game in history, Monopoly®[32].

We tend to rewrite the histories of technological innovation, making myths about someone who had a great idea that changed the world when, in reality, that person was usually not the inventor at all but the person who knew how to exploit the idea and bring it to scale. Such was the case with penicillin, where Alexander Fleming has been credited with an idea actually discovered and pioneered by a fellow chemist from the same laboratory, Howard Flory.

---

[32] For decades, Parker Brothers, the game's manufacturer's, peddled the story of how Darrow, an out-of-work salesman struggling to pay his rent in depression-era Philadelphia, devised the game while shivering in his freezing basement; how he carved little wooden buildings with frozen fingers, painting them green and red; printed fake money, and moulded miniature metal top-hats and cars; how the rights were bought in 1935; and how the game's popularity made him wealthy. This was an uplifting 'rags to riches' story. "There was only one problem," writes Mary Pilon in her book The Monopolists, "the story wasn't true".

It was a similar story with the light bulb. Invented by the British chemist Humphry Davy in the early 1800's, the concept of the electric light bulb spent nearly eighty years being passed from one hopeful physicist to another like an unwanted birthday present. In 1879, an American entrepreneur, Thomas Edison, finally figured out how to make an incandescent bulb that people would actually find useful enough to buy. At least, that's the story. But Edison did not actually invent the lightbulb any more than Alexander Fleming discovered penicillin. That honour fell to a one-time British rival and later business partner, Joseph Swann. Edison was the one, however, to bring the idea to a scale sufficient for the public to notice, and so came away with the credit.

Edison's achievement was not the lightbulb *per se*, but in putting together the electrical systems that contained all the elements necessary to make the light-bulb practical, safe, and economical to operate. He and his team at Menlo Park outside New York had to invent the parallel circuit; a durable glass bulb that wouldn't explode when hot; an improved dynamo; an electrical conducting grid; capacitors for maintaining constant voltage; safety fuses; insulating materials; light sockets, and on-off switches. Before Edison could make his millions, every one of these elements had to be developed into practical, reproducible components, and eventually fitted together into an integrated functioning whole.

The difference is that innovation takes an existing idea and applies it in a new way, whereas invention demands a radical departure from what is known. Innovation and invention, though talked of almost interchangeably, are not the same thing. Innovation entails using the stuff we already have in better ways; invention involves creating new stuff, new ideas, new machines, from scratch. Innovation is re-engineering, re-purposing, or re-inventing what people already know how to do.

The ATM was a prime example of the latter as it represented a radical departure from what was then considered 'normal', and then changed the way we behave. Being able to access our cash anywhere, anytime meant we didn't have to think about cash at all; cash was as near as the nearest ATM. We bought on impulse instead of planning every purchase ahead of time. All we had to do was punch in a few numbers and 'hey presto', out popped our cash. We didn't stop to think for a moment on how this miracle happened, or how we were interfacing with a machine on a whole new level to that of a kitchen dishwasher or a track-side vending machine that dispenses cans of warm fizzy drinks or melting bars of chocolate.

Times of rapid social transformation unleashed by technological genius are referred to in the 'tech' industries as "Gutenberg Moments". But the allusion is misleading. First, it took not one moment, but more than fifty years to turn printing with movable type into a flourishing business concern and even longer for Gutenberg's printing press to change the late mediaeval world. And, just as importantly, he wasn't alone: like Edison and Marconi hundreds of years later, he was well connected and had business partners. For centuries we've clung to the romantic view of Gutenberg, the lone genius. But a closer read of history reveals him as one in a long line of innovators whose success relied more on collaborative teamwork than on vision, imagination, or bloody-mindedness. Without a team behind him, Guttenberg's effort to complete his invention and bring it to the world would not have succeeded.

It was much the same with the ATM. Its inventor was well connected in banking circles and, through his employer, De La Rue, had access to more or less unlimited finance and engineering talent. The machine's major components, especially the cheque reader and number recognition technology, not only already existed, but De La Rue already held the patents for each. All that was required was for these pre-existing technologies to be bolted together in a new way such that the idea would become, like Gutenberg's moveable type, greater than the sum of its parts.

Ideas are cheap. At least, according to Neil Rimer of Index Ventures, a venture capital firm, they are. What is needed, he says, "is not just the idea but the wherewithal to realise the idea ... it's more about mind-set than skill-set."

"Charismatic original thinkers that can attract a team, convince sceptics that a business opportunity awaits, that have the skills to lead, and who are brimming with ambition, conviction, dedication and passion will always be able to convince investors and customers to come along for the ride," he went on to say. This describes the early days of the ATM perfectly and was just as relevant fifty years ago as it is now.

But there was another magic ingredient behind the invention of the ATM. Some people refer to this ingredient as 'luck', but this fails to capture what happens when new and disruptive ideas make it from the drawing board to the board room. The more correct term is 'serendipity' which means something altogether different. Having been brought up on Kipling's *Plain Tales From The Raj*, Dad knew what serendipity was.

We think of serendipity as a happy accident; a chain of chance encounters when seemingly random events coincide. But it actually has a very different meaning.

In 1754, Horace Walpole, a well-connected English 'Man of Letters', outlined a Persian fairytale about three Princes from the Isle of Serendip who had extraordinary powers of observation that would get them out of sticky situations when travelling. This old tale, he suggested, contained a crucial idea about human genius: The Princes were always discovering things they were not in search of. Walpole's insight was that skill, rather than random strokes of good fortune, was somehow being brought to bear. He coined the term 'serendipity' as something people do, not just the consequence of things that happen to them.

It was not just that Dad was frustrated by his bank one morning in not being able to access his cash, it was that he worked for a company that printed banknotes as well as manufacturing the mechanical devices that counted and sorted them. He had started up Europe's first armoured trucking firm that distributed the stuff too. Although he was unaware of it at the time, he was also imagining a solution to a problem he didn't know existed ... that banks were trying to resolve their differences with the labour unions over Saturday working practices. If motivation is part of the serendipitous process, it's worth remembering that he was at the same time desperately trying to stop the subsidiary company he was running from going bankrupt.

According to the author Pagan Kennedy, when people unconsciously dredge the unknown they are engaging in a highly creative act. What an inventor discovers is sometimes a solution in search of a problem, but "is always an expression of him or herself." This describes Dad very well. The idea of the ATM was in him all along. In having lunch with Harold Darvill – a seminal moment, which we will come on to – he may have been in the right place at the right time, but he was capable of seeing patterns that others couldn't see. "Innovation isn't all hard work and dumb luck," she notes, "It's about paying attention." According to her, this is how we dream things up that change the world. A surprising number of the conveniences of modern life – smoke detectors, the microwave oven, and X-Ray imaging among them – were invented this way ... with someone stumbling upon a new method of doing things, or capitalising on a chance encounter.

The Wright brothers would have recognised the process as it must have been a little bit like the early days of flight: The design team – a rather sceptical bunch of bicycle mechanics – would have understood how wonderful it would be to fly but none of them would have been quite sure what an aeroplane should look like. It would need a power source, but what type? Should it be at the front or at the back? It would need wings, but what shape? It would need ailerons, but what size? Through an iterative process of trial and error, it eventually became clear what the physics could do. What was not known was how to bolt all the pieces together in a way that would lift the whole machine off the ground safely and reliably, and then replicate the process. As a businessman rather than an inventor, however, Dad got it the right way round. He knew what he wanted to achieve; it was just a question of working backwards to achieve it.

Dad was an entrepreneurial Scotsman with a quick, questioning mind. His father was a civil engineer who built things to last. Having arrived one mid-60's Saturday morning at his bank to find it closed, he became frustrated at not having the freedom to withdraw his own cash at a time and place of his choosing. In that moment, he saw society had a problem. But could there be a practical solution where none had existed before? His epiphany was actually the result of conscious, methodical planning in support of a confluence of ideas, experiences and opportunities … he didn't actually work out the practical solution until later that evening in his bathtub. This was the Eureka moment.

Arriving at the solution, though, required a form of 'matrix thinking' comparable to a three-dimensional game of chess. Such thought processes involve putting order into a rather chaotic process of spotting, incubating and re-combining ideas from many different times, places, and events, and then putting them all together in a way that no-one has thought of before. Applying this sort of logic to the development of the ATM, it could be said that this type of goal-oriented thinking helped shape a highly focused function: an automated machine capable of reliably and securely dispensing cash 24/7 without the need for a human teller.

By virtue of circumstance, he was the right man in the right place at the right time: He was the managing director of a firm that, among other things, produced banknote-counting devices and he had just started up Europe's first cash-in-transit armoured trucking company. The parent company, De La Rue, printed banknotes. By inventing a machine which dispensed money rather than just making it or moving it around, he brought the business full circle. The engineering achievement of the ATM did not come out of thin air, in other words, it came out of a logical and methodical thought process.

It's also true that having your mind prepared to exploit an opportunity is just as important as spotting one in the first place. The engineering frame of mind is systematic and especially adaptive towards producing useful and practical solutions. As Guru Madhavan put it in his book *Think Like An Engineer*, "The engineering mindset sees structure where there is none, and is adept at producing utility under constraints, and making considered trade-offs in the scheme of what's available, what's possible, what's desirable, and what the limits are."

But something else was going on; something that in the age before intellectual property theft we used to call 'reverse engineering', a concept of 'backward design' where the desired outcome is pre-imagined and the engineering is applied in reverse to achieve that goal. For De La Rue, the automated cash-dispenser represented little more than a novel way of bolting together already existing mechanical parts rather than the conceptual paradigm shift it really was. Even when the orders began to flow, they never really got the point and, thinking they could not compete with US manufacturing prowess in those post-war years, were content to revert to the OEM model[33] and focus on supplying the components under licence, leaving the business altogether some fifteen years later.

In imagining, creating and refining the systems and mechanical processes of the ATM – including security, accountability, data protection, and resupply – Dad was working entirely backwards to form a framework for what we now call "telematics" … a system of systems that was eventually to unite computing, telecommunication, logistics and encryption technologies.

But he couldn't conceive at the time what disruption this would cause, nor that such disruption would be his legacy; he just wanted to get hold of his money at a time and place of his choosing, not that of his bank manager.

The ATM became the icon we didn't know we needed.

The operating principles of the ATM are based upon reliability. From day one, customers had to trust the machine. This meant it had to work and work every time. It didn't just have to provide the correct amount of money to the owner of that money and nobody else, but it had to prove it had done so by leaving an audit trail. Being exposed to the elements, it also had to work equally well in heat-waves and blizzards. Realising this, Dad ensured that the ATM was always over-engineered and that redundancies were built in from the start … 'redundancy', in engineering terms, being the duplication of critical components.

It didn't work very well at the start, however, and even the inaugural opening event at Barclay's Enfield bank branch in north London had to be faked, with Dad later confiding that he himself had to manually dispense the first ever packet of ATM cash to the

---

[33] The 'Original Equipment Manufacturer' is a company whose products are used as components in the products of another company.

actor Reg Varney[34] who was waiting on the street outside in front of the world's TV cameras to accept it.

"Being a British inauguration, it was a low-key event compared to Tokyo," Dad once confided during an industry dinner in 2008. "We opened Japan's first ATM for Mitsui Bank in the Ginza district in front of a crowd of over 10,000 people. This brought the place to a standstill, largely because of the large neon sign 15 floors in height which said 'get your cash here now!' which the Japanese public interpreted to mean 'get your *free* money here now!' It was absolute chaos."

Early testing had to be conducted in cold and wet conditions as the machine was to be mounted on the exterior walls of bank branches across the south-east of England in the first instance. The location had to be kept secret for fear of industrial espionage. Being almost perpetually cold and wet, Scotland was chosen as one of the ideal candidate locations. But this was quickly ruled out on two grounds: First, Dad's father-in-law, who just happened to be chairman of The Royal Bank of Scotland at the time, had already told his son-in-law quite plainly that perceptions of nepotism could not be tolerated, and that his bank could therefore not be involved with the cash machine project in any way[35].

There was a serious second factor to consider too, which was that it was by no means clear that automation would be accepted by the British public. There was a very real fear that the machines would be vandalised by those who feared – then, as now – being rendered redundant by automation. As it happened, the first six prototypes were repeatedly sabotaged by having honey poured

---

[34] Reg Varney, star of the BBC's hit TV series *On the Buses* was the best-known TV star of his day in the UK.
[35] Dad took the ATM idea to the rival Bank of Scotland instead, where the concept was embraced with open arms. Not, perhaps, the best way to endear yourself to your new father-in-law, but pragmatic business.

into the delivery tray, and for a while had to be guarded by the police.

Instead, on the basis that it was not only cold and wet, but that its good burghers were unlikely to communicate too readily with nosy journalists, Zurich in Switzerland was chosen as the proving ground. Everything went well enough through the winter months of 1966/67, except for one persistent malfunction: Every once in a while, the machine turned itself on without anyone asking it to, and proceeded to spew Swiss Francs all over the pavement. One such test – which Dad and the De La Rue executive team had flown over to observe – ended up with them all scurrying about in their pin-stripe suits on their hands and knees in the snow trying to stop a blizzard from blowing the notes down the street. Some of these notes were collected by Swiss passers-by who handed every note back. Such honesty so impressed Dad and the De La Rue directors that much of the ATM's future development – for instance, the testing of the radioactive ink, and blast-testing of the vault – was also later conducted in Switzerland.

After much head-scratching, it eventually dawned on the engineers wrestling to solve the problem of these seemingly random acts of generosity that they were initiated when passing tram-cars 'sparked' the overhead power cables. ATM's have been protected from such short-circuiting pulses ever since.

Another feature developed in Switzerland was the exterior cladding. Nobody really knew how secure an ATM needed to be; all they knew was that it would be exposed to the elements – including extremes of hot and cold, as well as rain and snow – and therefore to passing criminals. Being used to vaults, Barclays had already insisted that the ATM's interior vault be constructed to the same specifications as those in their basements. Using much the same logic, this was used for much of the exterior housing, too, with a sheet of brushed stainless steel used for the fascia. Dad and his engineering team thought this a bit over-the-top at the time, not least because they estimated it would take over six hours for thieves with blowtorches to cut through it. It would also considerably increase the weight of the machine, making it a challenge for those who had to install it into the exterior wall of a bank. Discussions with Barclays over what colour the machine should be were short, however, as the high tensile steel meant that it more or less had to be "any colour as long as it's brushed metal grey." But, inadvertently, and not by design, this engineering-led solution gave the ATM its rugged and dependable appearance which focus groups would later confirm contributed to people's trust in the machine.

The thickness of the steel gave rise to another problem which, with the UK's weather in mind, had not been anticipated: Winter temperatures in Zurich are very much colder than those found in the temperate climes of the UK, and rarely rise above freezing. The working parts began to freeze solid. To get round this problem, heat sensors were installed so that when the temperature fell to near zero, the metal facia heated up to stop ice from forming over the dispensing mechanism. This approach to solving problems through over-engineering was typical of the early years of the ATM's development. Despite the UK being nowhere near as cold as Switzerland, the heat sensors remained, and most ATMs manufactured today have them fitted still.

The ATM deserves its foundational myth.

Nicolai Tesla, the genius behind much of how electricity is used today, thought a lot about the nature of invention. In 1919, he wrote: "The progressive development of man is vitally dependent on invention. Its ultimate purpose is the complete mastery of mind over the material world." This pithy quote hits the nail on the head. Too often, the idea of invention is undervalued, both by those who seek financial reward for the investment of their time or money, as well as by society at large. The printing press, the lightbulb, and the internet are very different in what they do and how they do it, but all were inventions. We tend to take technological breakthroughs like these for granted, forgetting that our world would be a very different place today without them. The Wright Brothers bent quite a few airframes before finding one that actually got off the ground.

It was like this with the ATM. A man had an epiphany out shopping with his son one Saturday morning; thought about it in the bath that evening; discussed it with his sceptical wife the next day; went to work on the Monday and spent the rest of the day persuading all five of his equally sceptical employees – all of whom were facing imminent redundancy – that the idea was worth developing; received their first order a week later despite not knowing if the idea was even feasible; and then launched an unfamiliar technology on an unsuspecting public who, although initially bemused by the concept of instant gratification, went on to embrace it with such fervour that it is nowadays difficult to imagine life without one on every street corner.

Although no longer anchored in the romantic myth of the lone genius, ground-breaking stories from a bygone era like this scream out for a foundational truth; a story of vision, dedication, and engineering brilliance that eventually overcomes every hurdle to end up changing our world for the better, forever. National, scientific, political, and industrial icons are founded on such myths. Yet neither the ATM nor its inventor ever acquired mythical status … which is a little surprising given the ubiquity and longevity of the machine and the multi-billion dollar industry it went on to spawn.

Like the light bulb, the ATM was a failure at first. It has been argued that this was partly because the initial order was restricted to a single client, Barclays Bank; a monopolistic decision which, so the argument goes, hampered others from developing their equivalent machines. This takes a rather linear and myopic view of history, divorcing the story from the context of the time. Geniuses are important, but their achievements have to be seen in the context of the times they lived in, and the people with whom they collaborated.

In the mid-1960s there was little idea of how innovation should best be commercialised. Collaboration and sharing – what we would today call 'open-sourcing' – were quite normal. But running off with someone else's intellectual property, as the National Westminster Bank (now part of the Royal Bank of Scotland) did with the ATM, was considered as unethical then as it would be now. My Dad used to say that "The days of the 'word-is-my-bond' handshake ended with the ATM's arrival."

The debate over whether innovation is better served by sharing intellectual property or by protecting it has become more heated since the unveiling of the world's first working cash dispenser in those far-off, black and white days of 1967. But the nature of invention has been ever thus.

The days of the 'word-is-my-bond' business handshake ended with the ATM's arrival.

As with any other complex technology, cash-dispensing machines resulted from a long sequence of innovation, and not just by one company. June 1967 saw the arrival of the much-celebrated De La Rue automatic cash system known as DACS. Within weeks of its debut, two other independently engineered devices were unveiled. The first was the Bankomat which was installed in Uppsala, Sweden on 6 July. While the DACS was activated by a paper token, the Bankomat used a plastic card with an encoded serial number which was read by an optical device.

Also that month, the Chubb MD 2 was launched in the UK. This machine was activated upon insertion of a plastic card with perforated holes; a variation of the PIN concept, but essentially the same idea.

A third British team, this time from the Midland Bank (today HSBC) conspired with a small engineering firm called Speytec to come up with yet another version. Being the second biggest clearing bank in the UK, the Midland team were members of the original De La Rue-Barclays development group. They soon spun off, however, to develop their own adaptation. This version used a plastic card with a magnetic stripe embedded in it that stored a secret six-digit PIN which was returned to the customer after the transaction. This gave rise to a number of security challenges which the other two versions had not had to face. One of these was that the number of times the card could be used had to be restricted, a challenge eventually solved with the help of the National Physics Laboratory. As with the four-digit PIN concept, this team was well aware of De La Rue's 'three strikes and you're out' feature. Legal wranglings over this and related issues meant that the design couldn't be patented until September 1969.

Nevertheless, the Speytec machine, now with support from the Detroit-based Burroughs Corporation, went on to be deployed in large numbers, not just by the British savings banks, but in the US. The Chubb MD2 and Swedish Bankomats were highly successful, too, and became the machines of choice for many banks across Europe, in Canada, and even into the Soviet Union.

These early devices were not developed independently of each other, but were the product of collaboration between bankers and engineers in the UK, Sweden, and Japan. In the early days of development, the 'trade secrets' that made the machine work were shared amongst them all. As a boy, I would regularly be introduced to visiting Japanese businessmen who would stay overnight at our house outside Cobham, one of whom, in 1966, was so impressed that I was learning judo at school that he bought me a samurai helmet as a gift. I have it on my desk at home still.

The original ATM was 'cobbled together' using parts that already existed.

As with most inventions, the early days of the ATM were plagued with operating 'hiccups' and the rectification of design flaws. The first cash dispenser, the De La Rue DACS, was, quite literally, cobbled together using parts that already existed. When a single component broke, the entire machine had to be dismantled, a process that could take days if not weeks. Operation was hardly ideal from the customer's perspective either; one-time-use vouchers either had to be purchased from a teller before they were redeemable, all plastic tokens had to be returned by post for re-use. Withdrawals were limited to a single transaction per day and were only possible at the customer's own branch. It took years before it was possible to access your money from another branch, and even then it had to be from a branch of the parent bank. Managers at all levels within the main commercial banks thought the idea of the ATM would never catch on. "People," they would tell themselves, "like dealing with tellers." With each machine costing thousands of dollars (over $30,000 in today's money), they also thought they would never get a reasonable return on their investment despite the fact that, in theory, fewer staff would need to be employed.

The early devices worked as stand-alone units, and most ate the activation token rather than returning it directly to the customer. In this, it was operating more like a chocolate-bar dispenser or self-service petrol pump than a modern ATM. Customers eventually began to enjoy the convenience brought to them by this early automation, though, despite the significant technical shortcomings which hobbled their proliferation. And they were, of course, limited to only one function … that of dispensing cash when bank branches were closed. Nevertheless, as cash dispensing is one of the highest priority tasks of a bank cashier or teller, most of those in the cash management industry agree that this machine deserved its acronym, the Automated Teller Machine.

In a letter to Dominic Hirsch of Retail Banking Research, Dad explained the early days of the ATM's development from his perspective. He had never outlined what really went on in such detail before:

"In the spring of 1965, I had a very personal problem of not being able to cash a cheque on a Saturday, arriving at my bank in Cobham High Street, near where we then lived, exactly one minute after it closed. Instead, I had to persuade my local garage to cash my cheque. Lying in my bath that evening, I thought there must be some way of delivering money automatically through a hole in the bank's wall, around the clock. There was a way of depositing money all right, the all-night safe deposit box, but no way of obtaining money.

As the bath grew colder, I envisaged something like a deluxe chocolate bar dispenser unit of the type I sometimes used while waiting for a delayed train at the station on my morning commute to work. It needed some form of cheque, some personal input for authorisation, and an audit trail. That much I knew.

Everybody – at least those with bank accounts – used cheques in those days to withdraw cash; a time-consuming process which entailed standing in line at your branch, usually during your lunch hour. De La Rue was the country's main printers of cheques, but my bit of the company had yet to perfect the paddle-wheel note counter. The only option, therefore, was to pre-wrap a standard amount of money – I thought £10 was enough for anyone's week-end spending needs – in a 'brick', much like a chocolate bar, and exchange it for a specially printed secure cheque. This might prove the audit trail, but I couldn't for the life of me work out how to tie the owner of the cheque to the machine so that stolen cheques couldn't be used.

Some form of personal identification number was clearly needed, as anything more fancy, such as fingerprint or voice recognition, did not exist except in sci-fi movies. As I levered myself out of the bath, I thought a six figure number would do it. My Army number was a six figure number, and, like any former soldier, I could still remember that. I went downstairs, and outlined the idea to my wife, Caroline. She listened, said she couldn't understand why anyone needed to obtain cash out of banking hours, and then said, 'anyway, I can't remember more than four numbers at a time.'

I mused about this through the rest of the week-end, while mowing the lawn and teaching the children, then aged five, eight, and ten, how to do parachute rolls. 'Green light on, stand at the door … Red light on, GO!' One after the other, they jumped out of the willow tree, bending their knees and rolling away into the long grass. They never seemed to tire of this, but I was distracted.

First thing on arrival at the office in Regent Street in London on Monday morning, I gathered the team – all three of us – round the table and outlined the problem[36]. The bottom line was, could we invent a stand-alone device that would deliver cash through the wall of a bank? We discussed the

[36] Speech given by John S-B at an ATMIA conference in Florida, 22nd February 2007

security options: Carbon-14 marking of cheques – we had been working on something similar involving Geiger counters with a school chum of mine, Dan Stanley, who was Chairman of Pye, the electronics company, so knew that this might be feasible – retinal scans, fingerprint, signature, and voice recognition, that kind of thing. I had recently had cause to see the Foreign Office's filing system, and knew that at that time they punched sequenced holes in cards and then shone a light through the stack to see if any card fitted the criteria they were searching for. It was elegant, but not very accurate. We needed to be 100% accurate.

Having agreed the performance specifications – 'The machine must incorporate a personalised authorisation system; leave an audit trail in the dispenser; and deliver a standard batch of cash through a handle-activated drawer; the whole operation to take no more than 30 seconds' – I asked them to go away to think about it and re-convene in 48 hours.

On Wednesday morning, back-of-an-envelope calculations suggested that the idea was feasible, at least from a technical point of view, as long as we could control machine-readable codes and their related security ingredients. Everybody, though, agreed with Caroline's observation that customer demand did not seem to be there. One thing we did agree on, was the four-figure Personal Identification Number, from then on called the 'PIN'. Little did any of us realise it at the time, but Caroline had just set the global standard which is still in use today.

On Friday, wearing my other hat as Chairman of Security Express, I was due to host one of our regular monthly lunch meetings with one of the Clearing Bank general managers. We rotated the invitation, and, this time, it happened to be the turn of Barclays Bank. At the time, Barclays was the world's fourth largest bank, and we had a contract to move their money to and from over two thousand of their branches. Their Chief General Manager, Harold Darvill, turned up early at my suggestion, mostly because I wanted to discuss cash sorting challenges across the nation.

Emboldened by my second Pink Gin – and doubtless he with his – I drew him to one side as we were going in to lunch and asked for a minute-and-a-half of his time so that I might put our embryonic cash dispenser idea to him. John Finn, our financial controller, was with me, and said it took less than a minute for Harold to stop me in mid-flow and say, 'John, if you can do it, we will buy it. We'll finance the development costs and buy a few hundred devices if the prototypes work out.'

The team got to work that afternoon, but more in hope than expectation, fully thinking that Harold would call on Monday morning saying that he had re-considered over the week-end. Far from it. The next Monday afternoon, the Deputy Chief Executive of Barclays arrives in the bank's Rolls Royce at twenty minutes notice – I remember it clearly, because our office was so small, we couldn't find a parking space for it – and wouldn't leave my office until we had hammered out a letter-of-intent to proceed together. We agreed to develop six working prototypes, followed by 250 to be delivered in batches of 50. The specification and price was to be agreed between ourselves and the bank's Operations & Management Department. On leaving, and almost as an afterthought, he turned to me and said, 'John, do whatever it takes and spend as much as you need. Let me know if you have any problems.'[37]

Can you imagine? A Eureka moment in a bathtub one Saturday, an internal brainstorming on the Monday, a slightly

[37] Related verbally to James S-B on or about August 2009

wider feasibility meeting on the Wednesday, a sales pitch on the Friday, and a signed letter-of-intent and promise of a large order the following Monday. All that in little more than a week.

What I did not know at the time, was that, by pure chance, Harold Darvill had attended a meeting just the day before our lunch together with his Clearing Bank colleagues, where they had discussed Union demands to either shut banks altogether on Saturdays or pay the tellers substantially more to open up for cheque-cashing. In the early 1960's, banks in the UK were under pressure from employee trade unions for access to banking during normal working hours. Banks wanted to be closed on Saturday mornings shortening the work week. Since banks were already closing during the week at 3.30pm each day, a method had to be found to provide an acceptable level of service for customers who work during banking hours. No bank service on Saturdays and closing the bank two hours before most people leave work was starting to be unacceptable to modern work hours. An automatic cash dispenser was seen as the solution. What was to become known as the De La Rue Cash System, or DACS for short, gave the banks the answer they were seeking, not necessarily connected with efficiency, but certainly with customer service. This sort of lucky timing cannot be engineered in advance. Either way, De La Rue Instruments was reprieved in the nick of time.

The R&D team quickly expanded from its De La Rue Instruments members, to include the Head of Organisation and Method from Barclays, Ron Everett, and a small group from Pye Electronics in Cambridge. Pye had a good model shop, and extensive knowledge of working with Carbon-14. We had the experience of cheque printing and cash handling, whilst Barclays were experts at running internal banking systems.

It took us over two years to go from conception to first working prototype. One of the main stumbling blocks was obtaining approval from the UK Atomic Energy Agency (AEA) to use radioactivity in the public domain. But, as the use of a Geiger-Muller tube to read the principal cheque was key to our security approach, we had to accept their slow workings. This approach was eventually vindicated when the equivalent atomic agencies in Italy and Switzerland allowed the system to be used in their countries.

In practice, we ended with a special personalised cheque over-printed with Carbon-14 impregnated ink – a security feature used in some high denomination banknotes even today – and punched hole codes in the cheque which could be read by light emitting diodes (LEDs). The customer would open a telescopic drawer, lay the cheque on the positioning pins, close the drawer, type in a four-figure PIN, and then, if the LED readings and PIN matched, open the drawer again to find the cheque had been replaced with a packet of ten £1 notes."[38]

In the notes of his ATMIA/RBR after-dinner speech delivered on 14 April 2008 in London, Dad gave an overview of what happened next:

"I had the fun of selling the world's first cash dispenser, the DACS, to Barclays bank, the second to the National Bank of Switzerland in Zurich, the third to First Pennsylvania Bank in Philadelphia, and the fourth to Mitsui bank in Tokyo. All were firsts in their world.

---

[38] This section has been taken almost verbatim – the author has changed a few words for the sake of making the story flow more intelligibly – from John S-B's various written and oral testimonies.

In each case we did the pioneering job which in turn woke up the market to such an extent that I rapidly deduced we should stay only in that bit we knew the most about, the electro-mechanical dispensing mechanism. We, De La Rue, remained the world leader in these units until about 1982, a position which provided us with the income stream to get us into cash processing i.e the sorting, counting, and packaging of currency notes in circulation.

We got into this business shortly after the Bank of England's chief engineer told me at a lunch in 1969 to celebrate the commissioning of the bank's first and very own internal cash dispenser, that I should try to buy Crosfield Electronics as they were having problems with designing and making the world's first currency sorter. To cut a long story short, we did.

With this, the humble little company De La Rue Instruments, with its eight employees that I had parachuted in from the parent, Thomas De La Rue, turned into a £330 million per annum operation with 2,200 employees in 22 countries worldwide[39]. When the ATM came along, our turnover was £30,000 for the year, from which we made a modest profit of £3,000.

The US part of the story started in Florida in February 1969 where I was the first foreigner ever to be asked to address the American Bankers Association conference in Miami. Together with John Walker, my CEO, we gave a 15 minute presentation on the DACS to polite applause, no questions, and only 12 brochures taken away from the 2,000 provided. The general view in the bar afterwards was, 'who needs money out of hours anyway?'

---

[39] De La Rue eventually sold the business created by the ATM idea for £650 million in 2009.

71

Six weeks later, back in London, an unknown voice telephones from the airport saying he is head of operations at First Pennsylvania Bank and his chairman had flown him across the Atlantic at no notice with instructions to "Buy six of whatever those things are that Shepherd-Barron talked about in Miami." Not a very flattering story, perhaps, but it got us off the ground in the United States and, I believe, was instrumental in opening up the whole US market.

This was greatly helped by the desire of Citibank, then led by John Reed, to set the correct ATM specification for them to use internally, but also for use by their correspondent banks across the country. To this end, they set up a specialist company in San Francisco called Transaction Technology Inc., and where I introduced them to Earl Wearstler, later to become president and CEO of Diebold. The three of us working together came up with the most popular ATM design for use in America, assembled and installed by Diebold. I think it fair to claim that Citibank and Diebold did more to promote the operational concept of the ATM than any other company at the time.

It was around about then that we in De La Rue realised our pioneering days in the ATM were over, and that we should focus on being supplier of mechanical currency-handling systems to the trade instead. By 1982, over 70% of all ATMs in the world were using these devices, all made at a small factory at Portsmouth in the south of England ...

... Portsmouth, incidentally, is home to the Royal Navy's flagship, HMS Victory, made famous by Admiral Lord Nelson at the Battle of Trafalgar. The ship is open to the public, but certain areas such as Nelson's private cabin are strictly off-limits. It just so happened that the brother of our factory manager, who worked on the ship, had the key to this cabin.

Realising the effect having a 'tot' of rum from Nelson's own decanter in Nelson's own cabin would have on visitors, I would arrange for exclusive access, and it was in Nelson's cabin, shaking hands over the chart table where Nelson himself had planned the battle of Trafalgar, that most of our principal deals were cemented."

Patent records do not always prove who invented what, or when.

The conventional way to protect intellectual property is to patent it[40]. This gives inventors legal protection for their ideas: if others want to use it, they must pay for the privilege. The snag with this is that the idea first has to be published, making it easy for someone less law abiding to steal it. This someone could be a commercial company, an individual, or a country. So, in an age of rampant intellectual property theft, a lot of companies prefer to keep their most valuable ideas under wraps. Not every invention needs to be patented. It was much the same back in the mid-60's, especially where security protocols were involved … as they were with the ATM.

No one knows how many trade secrets companies keep, or how much they are worth. Some, like customer lists, are generated during day-to-day operations. Others are kept secret because patents typically last only 20 years. Had Coca-Cola patented its secret recipe back in 1886, it would have lost the rights to it long ago, and it would have lost its mystique – and its market – straight away. It's the same with pharmaceuticals. Once the formula is exposed, the chance of recouping the considerable research and development costs involved reduce considerably.

---

[40] A patent is a government license conferring sole right for a set period to exclude others from making, using, or selling an invention.

When Steve Jobs unveiled the i-Phone in 2007, he did more than change an industry. He did more than change the way we communicate, too; he changed the way we think about communicating. Apples' brilliant new device was the 'must have' item of its day – and to some extent still is – and represented a huge advance on the mobile phones that had gone before. It looked different. It felt good in the hand. It had more functions. Its interface was more intuitive. And it was faster. It also worked better. What the i-Phone was doing was taking what had gone before and refining it. As such, the i-Phone represented innovation at its finest.

The ATM which burst onto the Enfield high street nearly two decades earlier did not represent innovation, as there had been nothing like it before. It represented invention. It was new, it was useful, and what it did – although blindingly obvious with hindsight – was not obvious at all. In short, it fulfilled all the requirements for the granting of a patent. Except it was never patented. Why not?

To answer that question, it might be helpful to recall why patents exist in the first place. The system of Patent Protection was established as a trade-off that provides a public benefit, while allowing the patent holder to reap the commercial rewards of his or her investment in time and money. Essentially, the state agrees to grant a limited monopoly to an inventor in return for disclosing how the technology works. If, after technical review by a panel in the Patents Office, the invention is deemed novel, useful and non-obvious, the inventor is awarded 20 years of exclusivity.

A proliferation of patents harms the public in three ways. First, it means that companies will spend more time defending their patent in the courtroom than innovating or expanding market share. Second, it hampers follow-on improvements that take an existing technology and develop it further, often in ways the original inventor had never foreseen. And, third, it fuels many of the patent system's broader challenges, such a trolling (speculative lawsuits by patent holders who have no intention of actually making anything), defensive patenting (acquiring patents to pre-empt the risk of litigation) and innovation gridlock (the difficulty of combining multiple technologies to create a single new product because too many small patents are spread among too many small providers).

The original ATM consisted of over 300 different electro-mechanical parts. Some of these were patent protected in their own right. The 'paddle-wheel' roller that actually counted the cash, for example, had just been introduced to help tellers count paper cash at the bank counter automatically rather than laboriously by hand. Part of what made the concept of the ATM viable in those early days was that the same company, De La Rue, already held the patents on almost all of the working parts. It was just a matter of bolting them together in a new way to do something that until then nobody else had thought of doing.

To reveal how these parts worked together would, in itself, have been worthy of patent protection. But, as with the i-Phone, such a machine is more than the sum of its electro-mechanical parts, and its whole viability rested on the applications built into it, especially its anti-fraud systems. To demonstrate to an unscrupulous criminal underworld exactly how these systems worked would have rendered the whole idea obsolete before it even had a chance to earn the public's trust. It was a non-starter. And so, as a conscious decision, taken with full legal advice, filing for patent protection was never sought.

It is interesting to note that no one has developed a satisfactory statement of what constitutes an invention. Philosophically, this seems to follow from the fact that an invention is something which is found by reaching out into the unknown. Since an invention cannot be defined by describing something which is still unknown, the only alternative is to state what is not an invention. This is done in patent law by defining what is in the *prior art* ... which includes the entire body of knowledge from the beginning of time to the present. If an invention has been described in the prior art, a patent on that invention is not valid.

In the simplest terms, prior art is simply evidence that something similar has been done before a person applies for a patent. This involves 'proving' that this is the case – a process patent lawyers call 'interference' – normally by means of written evidence of publication or sale. Interference proceedings, however, are expensive and time consuming, and until they are completed, there is uncertainty as to who actually owns the patent rights. Moreover, most countries in the world abide by a first-to-file system where the patent rights are granted to the first party to file a patent application, regardless of whether they were the first inventor.

Sometimes – and this was the case with the ATM – written evidence can be kept secret i.e not made available to the public. Information kept secret, for instance, as a trade secret, is not usually considered as prior art, provided that employees and others with access to the information are under a non-disclosure obligation. However, the work undertaken by De La Rue and Barclays was under no such non-disclosure agreement. With such an obligation, the information is typically not regarded as prior art. Therefore, a patent may be granted on an invention, although someone else already knew of the invention.

Solutions are far from straightforward, with the main reason for favouring secrecy over patents being security. Elon Musk, for example, refuses to patent technologies developed at his Space-X rocket company for fear that foreign space agencies would simply pinch them.

However, keeping trade secrets is harder than it looks, especially as most such thefts involve insiders. These are typically employees or contractors who are given access to sensitive information, which they then 'borrow' – or, depending on your point of view, 'steal' – for commercial gain. This, indeed, appears to be exactly what National Westminster Bank did with the ATM. Here is Dad again:

"The original exploitation plan focused on Barclays Operations & Management Department helping us fine tune development to suit banking security requirements. They were happy for us to patent individual electro-mechanical components – De La Rue was heavily involved in security printing and development of similarly sensitive products, so knew full well how important patent protection was – but did not want us to patent the machine as a whole, particularly how the 4-figure PIN was to work. A long debate was held with Peter Etheridge, the De La Rue patents lawyer, and we reluctantly agreed, insisting that Barclays wrote up an accurate note of every weekly development meeting which we could use to prove prior art in the event of any subsequent patent infringement claim.

Whilst working officially only with Barclays, the general managers of all the other main banks quickly learned what was going on and wanted to be involved. Without signing any sort of non-disclosure agreement – we worked on 'my word is my bond' in those days – they were invited to join the meetings as observers. Lloyds Bank was happy for Barclays and De La Rue to shoulder the research and development costs and said they would come in after we had honoured our commitment to Barclays for six prototypes and a further fifty machines should they be proven to work. Midland Bank soon went it alone with Speytech-Burroughs, but they never really got going[41]. The National Westminster Bank (later NatWest), then the biggest bank in the country, were not happy to discover some months later that our contract with Barclays gave them exclusive rights to the first fifty working machines.

I found out exactly how unhappy they were when their General Manager, Tom MacMillan, rang me one morning to tell me in no uncertain terms that 'he was damned if he would play second fiddle to Barclays,' saying that, if we didn't re-negotiate the contract, he would take the idea to another party, including all our confidential ideas about audit trails, pre-packed banknotes, and, crucially, the PIN.

I warned him that, although he had not signed a Non-Disclosure Agreement, I thought his action 'a bit sharp (underhand).' But he guessed – correctly, as it turned out – that we would never sue because he was such a big customer of De La Rue's in every other way, and particularly of Security Express of which I was still chairman … we carried all their cash around the entire country; were now printing their Traveller's Cheques; and he had just become the biggest customer for our newly designed currency counting machines.

With NatWest being such a big customer – so big, in fact, that I was hauled up in front of our Chairman after he had been called by NatWest's Chairman, to justify terms of the deal with Barclays – I went over to Barclays in some embarrassment to see if it were possible to include NatWest in the deal? Not surprisingly, they laughed and would not give way. NatWest was their biggest competitor, and they were not going to give

---

[41] Midland Bank did, however, did go on to unveil the world's first mobile ATM (see photograph at the end of the book).

them an inch of what, if it worked, was clearly going to be a game-changer for their customers.

Shortly after this conversation, MacMillan took all our ideas to Smiths Industries' subsidiary, Kelvin Hughes, simultaneously telling them we were not patenting for security reasons. More galling still was that he used the same development team that had worked 'off the record' with our team.

The key component was the 4-figure PIN and its associated coding systems. With considerable knowledge of anti-fraud systems, we had gone out of our way not to patent these on the grounds that full patent disclosure would enable a potential fraudster to attack the system. Barclays were also particularly worried about this element.

So, the PIN coding system was given to Smiths Industries behind our backs. Uninhibited by fear of potential fraud, the designated project manager at Kelvin Hughes, James Goodfellow, filed for patent on the PIN, its associated coding protocols, and a number of the machine's other components, on behalf of his company. We obviously could have challenged his filing but, after much debate, chose not to do so for the same reasons we had not filed for patent protection in the first place.

Sometime in 1966 – about a year after development of DACS kicked off in earnest, and a year before the first machine was to be unveiled – Chubb, who had bought the patents from Smiths Industries in good faith, and who therefore had the right to exploit them, naturally proceeded against us for patent infringement. As they were making NatWest's cash dispensers at the time, they had every incentive to do so.

Our patents team, led by Peter Etheridge again, and backed by Barclays, met with Chubb's lawyers, and within less than half a day in a private meeting behind closed doors, was able to prove absolutely that De La Rue's work ante-dated every one of the patents, systems, and protocols Chubb now thought they owned, and that De La Rue could rightfully claim prior art. This thereby rendered Chubb's patents commercially valueless ... which meant that, while they were free to develop and manufacture their own machines, they could not stop De La Rue from doing so. The Chubb board formally agreed with their lawyer's recommendation some days later, and were polite enough at the same time to confirm De La Rue as the 'complete inventor of the ATM.' [42]

James Goodfellow, the engineer whose name was on the patents, along with others from Kelvin Hughes, later left the company, and nobody ever thought to tell him that what he thought were his ideas actually belonged to someone else all along. I feel rather sad in a way, because he must have spent many years thinking that he deserved more credit for what he honestly believed he had invented."[43]

Although neither De La Rue, Barclays, Chubb nor the UK Intellectual Property Office – formerly, the Patents Office – can now find written records of these meetings, the proof of the pudding, as they say, is in the eating. Had it been any other way, De La Rue could not have installed the first working DACS machine at the Barclays Bank branch in Enfield, North London, on June 27th 1967. Nor could the company have gone on to sell thousands of subsequent DACS machines and ATMs over the years. Chubb eventually left the ATM business altogether, leaving

[42] Transcript of interview between John S-B and James S-B, 26 December 2009
[43] E-mail exchange between John S-B, James Goodfellow, and Mike Lee of ATMIA, 28-29 January 2005

"many customers expressing satisfaction that De La Rue never published the coding system."[44]

Lloyds Bank, for example, bought DACS primarily because a patent had not been taken out. The same went for all the banks in Japan. In fact, the Japanese Ministry of Trade and Industry, the all-powerful MITI, gave De La Rue a 4.5% royalty on all Japanese manufactured machines for 10 years precisely for not having a patent, arguing that their banks were not put at risk thereby[45].

Dad was as surprised as everyone else when the Discovery Channel TV crew turned up on his doorstep, and until then he never considered himself in such a light. You see, from his typically down-to-earth perspective, he did not think he had invented anything particularly special in the ATM, far less anything to trigger a revolution. "What you have to understand," he once told me, "is that just being able to get your own money when you wanted it, not when the banks wanted you to want it, was revolution enough for me ... and all at no extra cost, either."

In fact, as a contribution to the history of financial services, he preferred to be known for his work in establishing and running Europe's first and largest armoured trucking business, Security Express. That's how we, his children, thought of it, too. We were too young to have bank accounts, so the idea of the cash machine soon palled. Security Express vans on the other hand could be seen, bursting with bullion – or so we assumed – in every high street of the land. Not only were these green vans with their gold logo more 'real' to us than the relatively rare ATM, but Dad had little 'Corgi' models made of them which we could play with and which gave us immense street-cred at school.

---

[44] Hand-written letter from John S-B to Ms Mclaren of IWC Media in reply to a letter from her on 26 June 2009
[45] Letter from John S-B to Mr Stephen Wiles at HM Treasury, 19 September 2005

A suit-wearing businessman, Dad's pivotal involvement in the ATM's development was no secret in the cash management industry. But it was only 40 years later and well into his retirement that the Discovery Channel 'discovered' him and took the notion further by declaring him to be "one of the top 20 inventors of all time." Yes, he might have revolutionised banking by bringing the world 24/7 cash, but he never sought fame or fortune and saw himself largely as a business pioneer, not as an inventor.

chapter four

security

: ~' / sə'kyürətē

noun
preventing the transfer of money or goods from a
person or their agent without permission; protecting
from force or intimidation

OVER THE EASTER WEEKEND OF 1983, an armed gang got into
the supposedly impregnable Security Express depot in East
London, neatly capturing the guards and holding them hostage.
In what is still one of the UK's largest-ever heists, the thieves
hauled away 5 tons of cash, worth £6 million (£26 million in
today's money). The gang made no mistakes and left behind not
a single clue[46]. The robbery was apparently masterminded by
Ronnie Knight, the former husband of famous UK actress Barbara
Windsor, who, coincidentally, lived in the same street as the man
who unveiled the world's first ATM, Reg Varney.

Attack, robbery, and fraud are an everyday reality, and for as
long as there has been cash, there have been criminals, con-artists,
counterfeiters, and latter-day bandit hackers trying to take our

---

[46] Ronnie Knight: Gotcha! - The Untold Story of Britain's Biggest Cash Robbery,
April 2003

money off us illegally. There are people out there, whole gangs of them, who, even now, are doing their best to part us from our money.

They do this in two ways: As in the Security Express example above, by physically stealing cash. Or by fraud, either in situ or remotely. There is a third way which will be discussed separately in the chapter on *Cash*, and that is to create new money by printing fake, counterfeit banknotes.

Physical attacks include the theft of the ATM itself and theft of the cash from anywhere in the cash cycle, including cash stored at cash centres, from cash-in-transit guards, of from customers in the ATM's vicinity. Fraud is more subtle in that it involves an intentional deception intended to induce another to part with something of value resulting in wrongful financial or personal gain by stealth. As far as an ATM is concerned, this includes a wide range of methods such as skimming, card theft and malware attacks amongst others.

To illustrate that the risks remain, two cash-in-transit security workers were attacked and seriously injured in a violent robbery by criminals wielding sledgehammers and crowbars as they carried money out of a supermarket to their waiting armoured van in Manchester, England in September 2016. Both men were beaten unconscious and were lucky to escape with their lives.

Traumatic as it must have been for the guards concerned, robbery of this old-fashioned *Butch Cassidy & the Sundance Kid* 'stick-em-up' kind is actually on the wane. As we will see, the introduction of smart cash cassettes and other sophisticated types of security device have rendered such efforts largely a waste of time and effort. Cyber banditry of a less deadly sort, on the other hand, has grown exponentially, with the expected shift of transactions to electronic forms of payment expected to see a correspondingly explosive increase in the already high likelihood of cyber-fraud. Globally, the true cost of such fraud is now estimated to amount to approximately US$3.7 trillion, whereas theft of physical currency totals a fraction of 1% of this amount. Either way, our cash is still vulnerable, whether it be a digit or a piece of paper.

Most ATM robberies occur at night between 7pm and 4am when the machines handle only 11% of total daily transactions but suffer 60% of the crime. Thieves are usually males under 25 years of age and work alone. There are several additional and distinct ATM robbery patterns, each of which presents a unique series of challenges. The majority of thefts are, however, one-off, opportunistic events, with over half involving robbery of the user immediately after authorising a withdrawal in what are known as 'transaction reversal attacks'. These are when the ATM 'times out' and begins to reverse a perfectly legitimate transaction by retrieving un-collected notes, only to have them grabbed by the thieves just as they are being swallowed. In other cases the thief forces the victim at gunpoint to go to an ATM to withdraw cash while on other occasions the victim is robbed of his or her ATM card, is coerced into revealing the PIN, while the thief then uses the card. In all these types of crime, the sums of money are relatively small. But more sophisticated and well-planned attacks can be very, very large.

In 2013, a worldwide gang of criminals stole a total of $45 million in the biggest ever fraud involving ATMs. They stole the money in a matter of hours by hacking their way into a bank's databases, eliminating withdrawal limits on pre-paid debit cards, and creating their own access codes before draining cash machines simultaneously around the globe. Co-conspirators loaded the data onto any plastic card with a magnetic stripe ... an old hotel key card or an expired credit card worked fine as long as it carried the account data and correct access codes. One security analyst described it at the time as a "virtual criminal flash mob."

While these hackers might have been sophisticated, using common criminals to withdraw cash from ATM machines shows that much cybercrime is still surprisingly low tech. "Run of the mill criminals are more common in cybercrime than you think," says Bill Stewart of Booz Allen Hamilton, a management and technology consulting firm.

The electronic tentacles of cyber-crime are finding their way into every corner of the globe, making digital money susceptible to all kinds of risk. In its 2015 Annual Report, even Barclays was forced to admit that it "faced a growing threat to our information … the integrity of our financial transactions … and to the availability of our services." And two years ago, Citigroup admitted that more than $2.7 million was stolen from 3,400 accounts during one such attack. And it's not just hacking and data theft; even central banks are not immune to fraudulent attack … as the $101 million attempted cyber-heist in Bangladesh in early 2016 recently demonstrated:

In a digital version of the bank heist depicted in the movie *Ocean's Eleven*, and in what would have been the largest bank robbery in history had it been pulled off, hackers breached the Bangladesh central bank's security system and then, masquerading as bank officials, sent a series of requests for the New York Federal Reserve to transfer large tranches of money to them. Since the correct protocols were followed, the funds were transmitted … wrongly, as it turns out.

According to banking officials in Bangladesh, the cyber criminals made an elementary mistake and were only stopped by a typing error in one of their transfer instructions … they misspelled the name of one of the intended recipients, writing "foundation" as "fandation". This very human error prompted an alert official at one of the routing banks to query the transaction, which automatically froze the payments. According to one rather red-faced Federal Reserve official at the time, "The payment instructions in question were fully authenticated by the SWIFT[47] messaging system in accordance with standard authentication protocols".

All this comment does is transfer our concerns about the security of our money onto another faceless link in the mega-money transaction chain ... which does nothing except make us even more anxious about where the future of digital transfers is headed. SWIFT is a private company which sells a supposedly super-secure system that banks use to authorise payment from one account to another. It is reputedly the 'Rolls-Royce' of payments networks. But three months after the attempted cyber-heist, SWIFT belatedly acknowledged that the thieves had tried to carry out similar crimes at other banks on its network by, as one New York Times journalist put it, "sneaking into the beating heart of the global banking system."

The admission that the attack was not a one-off event in a poorly regulated developing country the other side of the world, but part of a broader threat has raised questions about how securely mega-money – including our money – is being moved around. Financial security experts point out that, as with all other payments systems, the SWIFT system is only as safe as its weakest link. Sometimes, as with the SWIFT attack, the weakest link is to do with shortcuts in data encryption. Usually, however, authentication of the user is the culprit.

According to one FinTech company, Ripple, "the system is fractured and antiquated. The way it is set up, you cannot totally isolate problems in a place like Bangladesh from the whole network." This not only reflects the increasingly global and

---

[47] Society for Worldwide Interbank Financial Telecommunications

interconnected nature of finance, but also demonstrates the risk involved when many types of financial institution are transacting through a single system made up of a patchwork of banks and companies with varying levels of online protection.

According to Adrian Nish, head of the cyber-threat intelligence team at BAE Systems, a defence and security company, "Even if officials at the Bangladesh bank had employed the highest of security measures, the thieves displayed a level of skill, cunning and determination that could easily have penetrated a far more secure system." This reflects a growing sophistication among cyber-criminals, who for years have been trying to get at our money by breaching personal bank accounts and stealing credit card credentials. To ram home the gravity of the threat, Nish added "keeping them out over the long term is going to be really difficult."[48]

To make matters worse, the bad guys are winning the cyber arms race and are able to keep one step ahead of the law. According to the UK's National Crime Agency, cyber-criminals are more advanced than the police. "The accelerating pace of technology and criminal cyber capability development currently outpaces the UK's collective ability to respond," says the agency.

A common thread to most attacks is compromised or hijacked credentials that allow an attacker to pose as a legitimate entity. Either way, attackers are becoming more sophisticated and defences that people have relied on for some time are beginning to break down. Burgeoning rates of cyber-fraud resulting from this volatile mix means that payment providers should be looking to develop and adopt a stronger, less hackable alternative to the 50-year-old, four-digit PIN which is commonly used to authenticate the user's identity at an ATM. Being one-dimensional like passwords, it's clear they don't always provide

[48] Michael Corkery: New York Times, 1 May 2016

adequate protection even when we change them regularly ... which we tend not to do.

Cyber-defences that people have relied on for some time are now beginning to break down.

The ATM industry is at the forefront of this high-tech struggle of wits where, as the technology evolves, so crime adapts to circumvent each new security measure in a never-ending cycle of cat and mouse. Yet, as the criminal underworld has become more audacious and sophisticated, so have the ATM's security defences. This is because, in a bid to maintain customer loyalty and trust, the industry is constantly and exhaustively analysing the way the criminal mind works and looking at the different ways ATMs are attacked around the world. Based on these observations, a network of defences has been designed to guard against each type of attack.

Not too long ago, it took a group of hackers only 30 minutes to steal $9 million from 130 ATMs in 49 cities around the world. Europe-wide losses due to ATM fraud increased 13% in 2012 to reach €265 million according to figures provided by the UK Credit Cards Association. Over three-quarters of these losses were caused by card skimming attacks ... the most common type of which involves the wonderfully named 'deep insert skimming' variant[49].

Explosive attacks against ATMs are booming in Europe, according to the European ATM Security Team, rising 80% in the first half of 2016 against the same period a year earlier. Of the 492 attacks, gas was used in roughly four out of five (382), with the remaining one-fifth (110) involving solid explosives.

---

[49] Deep Insert Skimmers are wafer-thin fraud devices made to be hidden deep inside the card acceptance slot.

Overall, ATM-related physical attacks rose by 30% from 1,232 to 1,604 during the same period but actual cash losses only rose 3%, meaning that the amount of cash stolen per attack is falling dramatically. The average cash loss for a ram-raid[50] or burglary attack is estimated at €17,327 ($19,055); the average cash loss per explosive attack is €16,631 ($18,289) and the average cash loss for a robbery is €20,017 ($22,013).

ATM-related fraud attacks saw a 28% year-on-year increase during the first six months of 2016, with a total of 10,820 incidents reported. This was mainly driven by a three-fold increase in the sort of transaction reversal frauds mentioned earlier. Losses were up 12% over the same period, now amounting to €174 million ($191.3 million).

At the same time, there were approximately 20,000 skimming incidents in the US in 2013 with the average cost per incident approaching $50,000, equivalent to about $2,500 per ATM per year[51]. Since the US makes up about 20% of the global ATM market, worldwide ATM fraud could amount to as much a $5 billion annually.

ATM security is not just about protection, but detection.

As it's clearly no longer possible to separate the money from the machine, it's time we looked at how the ATM, its components, its contents, and its systems, work together to foil the bad guys. With more and more 'cash recycling' ATMs now accepting cash deposits and re-distributing those deposits to other consumers, it

---

[50] Ram-raiding refers to the physical removal of an ATM from its hole in the wall using extreme force and driving it away to be prised open later.
[51] The number of US ATMs compromised by criminals rose a staggering 546% in 2015 over the year before. This surge was probably due to criminals working to exploit mag stripe card vulnerability before EMV migration in the US reaches critical mass and thereby making skimming unprofitable (Source: FICO).

is no longer just a story of robbery prevention but also one of counterfeit banknote detection.

ATMs sourced solely with wholesale notes are unlikely to contain counterfeits as each and every note has already been authenticated by the bank which owns the machine or by the strictly regulated and licensed commercial provider during the sorting and cassette preparation process. Retail notes on the other hand are sourced from other customers when making a deposit or by a merchant replenishing the machine with takings from the till. Given that recycling is becoming more and more common, this leaves ATM operators with a challenge.

A recent 'code of conduct'[52] requires that all recycled notes are authenticated before they are dispensed through ATMs. This puts an enormous burden on the ATM operators as it means they are expected to be able to separate counterfeits from genuine notes in real time.

As stand-alone machines, ATMs can't combat crime themselves but they can – and must, by law – ensure that they don't unwittingly connive in a criminal act. Fifty years of technological evolution, all of it at the cutting edge of security management, has been brought to bear to ensure, to the extent possible, four specific aspects are safeguarded:

- The person attempting a transaction is who they say they are
- Any cash dispensed is genuine
- The vault cannot be breached by an unauthorised person
- Personal data is secure

---

[52] In the UK, the Payments Council operate the code on behalf of the Strategic Cash Group, whose members include the Bank of England, the Royal Mint, wholesale cash operators and representatives of retail banks and other cash industry stakeholders. The Code is also supported by the British Retail Consortium and the LINK network.

To achieve this, not only must the hardware be physically robust enough to withstand being blown up but the software has to protect all aspects of the financial transaction, including our personal data. This means that network security and audit trails cannot be compromised while all aspects of physical security in and around the machine, including attack detection, prevention and even pre-emption, have to be taken into account.

So, apart from software hacking, staff being attacked by criminals while in the process of transporting or replenishing the ATM, and physical removal of the machine by ram-raiding, to what types of attack is the ATM potentially most vulnerable?

Skimming: Despite card skimming in Europe significantly declining in 2011, it is still a significant form of ATM fraud – especially in the US where 'CHIP and PIN' (EMV) technology is still not widely deployed – and continues to be one of the major threats to ATM security, representing about 80% of all attacks against ATMs around the world. The technique 'skims' an unsuspecting user's card data which is used later to create a duplicate or cloned card. This is then matched with the PIN which has been secretly recorded at the same time. In a typical attack, a device looking exactly the same as the original is slipped over the card reader slot. This device contains a magnetic stripe reader and short-wave radio which transmits the captured data in real time to a nearby receiver.

For skimming to work, the fraudster must be able to match up the skimmed card's magnetic stripe data with the corresponding PIN. This is recorded as it is entered via a pinhole camera placed somewhere above the keypad, and is often disguised as one of those little mirrors that you use to see what is going on behind you ... which explains why it's so important to cover your hand when entering your PIN.

More commonly, the criminal just peers over the user's shoulder ... a technique that goes by the sophisticated title of 'shoulder surfing'. To complete a successful card cloning attack, the attacker needs the cardholder's PIN. The oldest way of doing this is simply to lean over the victim to watch the cardholder enters his or her PIN. This practice can also include installation of a hidden pinhole camera or use of long-range photography. Some of these hidden cameras use infra-red heat sensors to record not just the numbers pressed but in which order. And they can do this up to half an hour after PIN entry.

As was previously mentioned, deep insert skimming is a growing threat. Deep Insert Skimmers are different from typical skimmers because they are placed behind the shutter of a motorised card reader, and completely hidden from the consumer at the front of the ATM.

False Keypad: This is when criminals place a false but realistic keypad over the actual keypad. This false keypad allows the depressed keystrokes to be transmitted to both it and the real keypad underneath. This is usually so realistic that the user is normally completely unaware of the fake pad.

Fishing: ATM cards can be stolen by inserting hooks and probes into the card reader, causing cards to get stuck inside. Criminals return later with a "fishing" device to extract the cards. The criminals obtain the PIN either by shoulder surfing, persuading the customer to re-enter it while pretending to 'help', or, as with skimming, using small hidden cameras.

PIN Interception: Criminals capture PIN numbers as they're being sent from the keypad to the internal or host computer by tapping into the wires inside the ATM, or by remotely recording the electromagnetic radiation emitted by the ATM's internal wiring.

Trapping: Adhesive tape is inserted into the ATM, causing notes to get stuck inside. The criminals move in to remove the cash once the user has departed. The same applies to deposited envelopes.

False Transaction Reversal: Criminals defraud the financial institution by making the ATM believe a transaction was reversed, when it wasn't. Criminals wait for the ATM to begin retracting the notes, and then quickly pry open the door and grab the cash. Other criminals carefully remove some, but not all, of the notes presented. In both cases, the ATM believes it timed out and the transaction was reversed, while the criminals have stolen the cash.

False Front: A fake dispenser is placed on top of the ATM's dispenser opening. The ATM dispenses the cash into the false front, instead of to the customer who walks away in frustration thinking the machine is not working. The criminals return later to pick up the cash.

Having listed the threats, let's look now at how these scams have been defeated. A number of techniques are used, some of which are obvious to the user, some of which are not:

Fascia Design: A variety of different surfaces and protrusions make it difficult to add any kind of external device. The unique rounded shape of the fairly recently introduced bit of green flashing plastic around the card reader slot makes it almost impossible to install a skimmer, for example.

Anti-Skimming Device: You can't actually see the 'anti-skimming' unit behind the plastic exterior but you can feel the vibration in your card as it is slowly drawn into the card slot. The deliberate introduction of this 'jitter' – which, if you're not used to it, makes you think there's something wrong with the device – imparts an irregular motion which distorts the magnetic-stripe details on the back of the card making any unauthorised copy unreadable. Introduced ten years ago, 'jitter' is only effective on ATMs with motorised card readers – readers that pull the card in, read the data and then push the card out – and doesn't work on machines with 'dip' readers, in which the user manually inserts and withdraws the card.

Most ATM users will be used to introducing their cards 'vertically' i.e the thin edge first. A recently introduced anti-skimming measure, and one that is especially effective at countering deep insert skimming, involves inserting the card sideways (horizontally) instead.

This begs the question why fraudsters would bother skimming cards from ATMs in Europe, which were equipped to read data off the chip embedded in the cards issued by European banks long ago. The trouble is that virtually all 'chip & pin' cards still have the account data encoded in plain text on the magnetic stripe on the back of the card so that the cards can be used in the 50% or more of American ATMs that have not yet migrated to EMV.

Touch-Sensitive Screens: Increasingly, ATMs are using touch-sensitive digital screens. This allows PINs to be scrambled so, when entering your PIN, if someone is watching, they can't later use what they thought they saw you press as the position of each key on the screen changes from one user to the next. Nor can someone watching you easily figure out what your PIN is based on how you moved your hand or finger. It also renders use of an infra-red heat-sensitive camera obsolete.

Each time the PIN is entered, the numbers, colours, shapes, and symbols randomly change position on the screen. So, while you're pressing the same sequence of keys – on either the keypad or on the touch-sensitive screen itself – different keys would be pressed every time a withdrawal was made, making it impossible to see someone else's PIN without seeing the screen and keypad clearly. Called *Tri-Pin*, the idea came to inventor Glynn Reynolds after becoming a victim of ATM fraud. "It's a very simple but effective idea with a higher level of security than traditional PINs," he said. The system would avoid many current practices used by fraudster to steal PIN numbers, such as 'shoulder surfing' to see which keys are being pressed.

Near-Field Jamming: A multi-frequency jamming module within the ATM blocks 'listening' devices from capturing your card data by emitting electromagnetic 'noise' at such intensity that a magnetic stripe reader head in a skimmer device within range only captures scrambled data.

Card Reader Sensor and Security Bolt: Anti-Fishing mechanisms contain a motion sensor that detects the presence of a trapping or fishing device. When a problem is detected, the card reader raises a bolt that prevents the removal of the card and reports the activity.

Card Reader Monitored Gate and Protective Shutter: The card reader gate inside the ATM remains completely closed when not in use. At the beginning of a transaction, a protective shutter shields the inside of the card reader by only allowing the card to be brought partially inside. The card must pass a security verification check before the card reader will open its shutter and allow the card to be fully inserted.

Recessed Keypad and Display: By tilting the keypad, recessing it deep into the body of the ATM, or covering it with a flexible plastic shield, the user's body will naturally block the PIN pad. Similarly, a recessed and/or tilted screen using polarised glass makes it difficult for anyone other than the user to read what's displayed on the screen.

Dual-Locking Gate: The dispensing gate is engineered to form a tight seal as it closes. The door has a strong, robust locking mechanism with dual-locking features, making it difficult for criminals to open and gain access.

External Cameras: High-definition cameras continually monitor movement within the immediate vicinity of the ATM. When aberrant activity – for example, when fitting a skimming device – or the presence of more than one user is detected, the system temporarily closes down. Both overt and covert cameras are used. A single prominent camera is usually enough to deter criminals by making them aware that their actions are being recorded. Almost invisible digital pinhole cameras also record every transaction thereby making identification of anyone engaged in criminal activity much more likely. All these cameras send automatic alarms if the lens is covered or tampered with.

Most ATMs incorporate cameras capable of recording a human face and a vehicle number-plate hundreds of metres away. As anyone familiar with the popular TV series *Breaking Bad* will know, such information is often used for the successful tracing and conviction of criminals.

Anchoring: Installed ATMs have rock-solid island-anchoring systems that can withstand 72,000 pounds of force. An interior steel stabiliser adds extra reinforcement. And dual action live bolts secure the dispenser and depository within the safe.

Drilling Detection: Penetration mats line the ATM's interior in order to detect when holes are being drilled into the machine. Being linked to seismic sensors it is possible to differentiate between the vibrations caused by drilling and those caused by minor earthquakes or vibrations from unusually heavy passing traffic.

Electronic Locks: These use multiple passwords, giving different levels of access to different personnel. Multiple combinations require more than one person to open the safe. Remote access requires a signal from a remote facility to open the safe. Intelligent monitoring prevents the safe from being opened during unauthorised time periods, and detects unauthorised access based on predetermined usage profiles. Authentication identifiers, random number generators, and installed digital certificates are all used to prevent possible fraud by maintenance operators ... who may be being coerced.

Geo-fencing: This technique adds smartphone geo-location to the panoply of counter-measures described above. With this system, the card issuer will simply check the location of your smartphone to see if you're in the same place as your card. This 'geo-fencing' has the ability to stop an ATM transaction unless your smartphone is nearby but in reality takes a range of other security features into account before doing so.

One-Time-Use PINs: A refinement of the 'geo-fencing' system could also send you a one-time-use PIN via SMS message to your smartphone. This system is used widely in Turkey, for example.

Card Lock: Another way of ensuring your ATM card is not being used by someone other than you and put you in control of your own risk is to lock it while not in use. Such cards lie dormant for all but a few minutes of each day, so why not keep it in locked mode during these times, and unlock when you want to use it? It only takes two button presses on you smartphone.

Anti-Ram Raiding: When unauthorised removal is tried in such a violent way, the more modern machines are able to fill with foam which sets solid in seconds making removal of the ATM's inner workings all but impossible.

ATMs are packed with 'secret' smart sensors.

ATMs use a wide array of 'intelligent' sensors to counteract theft and monitor their own health. These 'secret' sensors include a number of specialised types designed to detect security breaches, some of them before they have even taken place. These include:

Door Sensors: These monitor the status of the vault door, and not only record when unauthorised access is being attempted, but sounds an alarm in the central control room as it is happening. In practice, the alarm will already have been raised as even opening the outer door of an ATM either outside pre-set times or by unauthorised persons will have been noticed.

Motion Sensors: ATMs have multiple motion sensors which are capable of detecting unusual movement in three dimensions. These are the tiny electro-magnetic accelerometers used in smartphones to orient the screen, and are capable of differentiating between an attack, an earth tremor – important in somewhere like Japan, which experiences tremors almost every day – and passing traffic.

Thermal Sensors: Micro-thermocouples are fitted to ATMs for two reasons: To stop the mechanical parts from freezing up in sub-zero temperatures – it would be a pity to struggle through miles of snow-drifts in colder countries like Norway or Nepal to find the keypad frozen solid – heat pads are automatically turned on whenever the outside air temperature falls below zero.

ATMs have also been subjected from time to time to successful 'freezing' attacks where liquid nitrogen is injected into the interior of the machine to make the metal parts more brittle. Similarly, use of an oxy-acetylene torch on the vault will raise the temperature inside. In both cases, neutralising foam is released which instantly turns solid when mixed with oxygen or nitrogen.

Gas Sensors: ATMs have also been subjected to other forms of gas attack, carbon dioxide being the favourite.

Magnetic Sensors: Curious as it may sound, each ATM has its own magnetic signature. This is something to do with where it is located as well as how and when it was manufactured. When this field is disrupted – when a metallic device is fitted on the inside or outside of the casing, for example – the machine locks down.

Microwave Sensors: Criminals have attempted to disable the internal electrical systems of many ATMs by subjecting them to pulses of microwave radiation. This is the method used in the movie *Ocean's Twelve* where George Clooney and his gang disable Al Pacino's casino security systems using a pulse generator stolen from a nearby research laboratory. The theory is based on the 'electro-magnetic pulse' generated by a nuclear blast which melts the wiring of anything electrical within a certain range. To an extent, it works, inasmuch that the ATM's wiring is affected, but not before the machine, having detected the pulse, closes down.

Even the most sophisticated array of so-called 'smart' sensors and the algorithms which combine and compute their various inputs would be useless if they were not linked in some way to an alarm system to alert virtual or human operators that something abnormal is going on, and that action might be required.

At the most basic level, detection of a physical attack by any one of the sensors mentioned above will trigger local and remote alarm systems. Usually, these are only audible to the remote security management team, who, once alerted, will confirm that the alarm is not false by cross-checking with on-site video and audio input before responding themselves and/or calling on the local police service. Increasingly, these systems are virtual i.e require no human analysis.

Electro-mechanical counter-measures can nowadays detect any attempt to tamper with the ATM's physical structure, inside or out. This includes attempts to disable or otherwise interfere with the ATM's jamming signal. The fitting of unauthorised devices will thus be detected, often without the criminals involved knowing that they have been 'rumbled'. With multiple digital pinhole cameras recording every transaction in high-definition, and streaming data to a remote backup, their images – along with those of people and vehicles across the street – will also have been captured … all of which increases the chances of being caught.

Today's ATM is a sophisticated machine supported by highly complex systems.

Despite the average European and American using an ATM 75 times every year, how it works remains something of a mystery. A card goes in, a few buttons are punched, there's some mysterious whirring and clunking, and, like magic, our money pops out. So trusting in this process are we that only rarely do we stop long enough to count it. The notion that something can go wrong rarely crosses our minds. Partly, this is because ATMs seem to carry out their basic function of dispensing and accepting cash faultlessly time after time. We trust them to work. And we trust the people that own them to make sure they work and keep our money safe. And they do. But how they do this has, until now, been a bit of a secret.

ATMs are not photocopiers ... paper trays cannot be refilled and jams cannot be rectified by the next random user. In fact, so sophisticated are a modern ATM's inner workings – a modern ATM has over 11,000 working parts – that even a bank branch employee has practically nothing to do with an ATM terminal, apart from installing paper, collecting retained bankcards, and resolving minor system incidents.

This highly complicated piece of engineering requires a vast and complex array of service support operations to keep it fully functional day and night, winter and summer. To ensure it continues to meet the evolving needs of consumers requires the brightest minds in business intelligence and financial technology.

But, before knowing how an ATM works, it would be useful to be reminded of some of the functions it can be expected to perform ... bearing in mind that not all ATMs need to perform all possible functions all the time.

The functions of an ATM have evolved since it first emerged in 1967 to dispense pre-packed amounts of cash, and now provide over two hundred options ... from the basic withdrawal, to making deposits, conducting foreign exchange, cashing cheques, paying utility bills, making transfers, ordering flowers, making mortgage payments, repaying loans, and even, somewhat morbidly, purchasing funeral plans.

For an ATM to perform these many varied functions, some basic components are needed. Some you can see, but most you can't. Those you can see include the screen display, PIN pad, card reader, and biometric scanner.

The bits you can't see include the control process unit (CPU), deposit and recycling modules, note picker, dispenser, and cash cassettes.

With more and more ATMs now accepting cash deposits and cheques, consideration has to be given to counting the notes (and sometimes, coins) correctly. As usual with such a complicated machine, this is not as easy as it looks. For a start, the notes don't just have to be counted correctly, they have to be recognised for what they are by currency as well as by denomination. This means forgeries have to be detected and separated. And finally, the deposited notes have to be sorted and stored. This is all done by a single stand-alone component, a banknote counter. Amongst other things, high-end counters can count 1,500 notes per minute regardless of the number inserted or the way they are orientated when inserted (vertical, horizontal, back to front, or upside down); recognise eight different currencies, each in six different denominations; detect counterfeits by verifying compliance with ten separate 'security devices'; and sort and store by currency and denomination.

Because the notes have been verified at point of deposit, the sum can be instantly credited to the customer's account as cleared funds. Sums deposited the old way – i.e in an envelope – take three or four days to clear as a human teller has to physically open the envelope, count the cash, and make the entry.

Counterfeit notes are not returned but diverted to a separate sealed bin which can only be opened by security experts, not by bank staff. Those notes which are accepted are either deposited into a single bin for later sorting by bank staff, or immediately re-cycled within the machine for later disbursement. Even with high-end ATMs, this is usually limited to three currencies and two denominations per currency, meaning a total of six automatically replenishing cassettes being housed within the ATM. This is no small order and represents the cutting edge of ATM technology.

Cash Recyclers allow deposited cash to be recycled and dispensed to subsequent customers. In 2004, Shinsei Bank in Tokyo went live with the world's first cash recycling ATM application. Recycling ATMs require fewer visits for cash replenishment because the deposited cash can be used straight away for the next customer's withdrawal. Fewer replenishment visits mean lower service costs.

Cash recycling systems are designed using currency recognition and verification technologies designed to improve the accuracy of cash handling. ATMs take over the time consuming and labour-intensive task of counting and verifying banknotes. As no ATM could possibly recognise every denomination of every currency in the world, far less sort and store it all, the sorting technology has to be customised to the currencies most usually passing through the bank. According to the bank ABN Amro, there is a mismatch in European ATMs of 30% between sums deposited and withdrawals made, with €100 being the most popular for withdrawal, but only €50 for deposit.

A hole-in-the-wall ATM typically holds between four and six plastic cassettes filled with cash. When a cassette is inserted, magnetic sensors on the side of each identifies the type of cassette and its contents, thereby allowing the pick module to select the proper one.

The pick module lies at the heart of the ATM. Having identified the correct cash cassette, the notes are 'picked' by small, air-activated suction cups which, when the suction pump is interrupted, drop the note into a series of rollers to be gripped and whisked up to the dispenser. This component is the whirring noise you sometimes hear when picked banknotes are being counted. It can dispense these at the rate of ten per second.

Smart cassettes have made cash almost unstealable.

ATMs hold cash. As long as they continue to do so, criminals will attempt to steal from them. Given sufficient time, tools and determination, they eventually figure out how to overcome the security counter-measures just outlined. So, is there anything further that can be taken to make this not worth their while?

The days of cash being counted by hand and inserted note by note into a spring-loaded drawer are long gone, at least for bank-operated ATMs. Instead, 'bricks' of single-denomination, single-currency notes are packed into highly sophisticated, but really quite flimsy, plastic cassettes. These cassettes are getting smarter and smarter, and incorporate ever-more effective ways of foiling would-be criminals. This is because such systems provide complete 'end-to-end' protection for cash from the moment it is packed, when in transit and whilst replenishing the ATM. Cash is constantly in a safe micro-environment.

Adopting the end-to-end approach, there is absolutely no need to open cassettes at any point of the delivery-collection process, so the most vulnerable parts of the cash handing cycle are eliminated. Risk is reduced for the financial institution and its customers as well as for the brave guys driving the armoured cars and carrying the cassettes across the pavement. ATM Cassettes can only be opened using a special device with a unique one-time-code. The person carrying it at the time of an attempted robbery either doesn't know the code, or doesn't have the device, or both. So, it's no good holding a gun to their head; they simply won't know how to open the wretched thing.

Cassettes can also be fitted with a tracking unit using a combination of GPS, GSM and GPRS technology, so that the cash management companies know where the cassette is at any moment in time. This is exactly the same technology used by logistics companies such as DHL to track packages through their system. Cassettes using this system are pre-programmed to specific locations and specific times, so that if the cassette is 'lost' (i.e stolen), tampered with by anyone – including by the person carrying it across the pavement – or inserted into the wrong machine, the money it contains is instantly destroyed.

When fitted with sabotage switches, the cassette's contents can also be configured to self-destruct when it is removed from an ATM or CiT vehicle at the wrong time, after input of the wrong de-activation code – which, like the 'mother' ATM itself, allows three attempts – or if someone tries to melt it, freeze it, or otherwise try to force it open.

These so-called 'smart' cassettes are also connected wirelessly to a central controlling unit installed inside the ATM. This controller receives inputs from several external sensors – a gas detector, seismic sensor, camera, and shutter intrusion detector, for example – each of which can trigger the self-destruct response.

As if these 'smart' cassettes weren't enough of a foil, they also contain accelerometers of the type used to orient the screen on our smartphones. If these detect the sort of violent movement caused by an explosion or road-traffic accident, they, too, instantly trigger the destruction of the notes.

ATM cassettes are also equipped with on-board smart chips, which store and carry a variety of information, such as how many notes are contained in the cassette, the currency, the denomination, access history, and error codes. The devices also track which bank employee loaded cash into the cassettes as well as who installed and removed the cassettes from the machine, and when.

So much for the cassette. But what about its contents? One way to deter would-be bandits from making off with someone else's cash is to irreversibly deface or destroy the banknotes before they have a chance of being used. This implies that the physical properties of the banknote must be altered if any attempt is made to steal it, not just altering the way it looks. Until recently, the preferred method was to spray the 'brick' of notes with indelible dye. But this tends to discolour only the edges of the note, especially if the 'brick' has been newly printed, which most retailers fail to spot over the counter and therefore means they remain useable ... and, even if they did, stained banknotes only indicate they might have been stolen and, in most countries, remain legal tender.

But, with criminals apparently fixated as they were in the days of the UK's 'Great Train Robbery' on the prize of stealing banknotes, there is one new technology which is proving to be a game changer. This technology – or rather, an old technology newly applied – ensures that the contents of a cash cassette are left worthless. Through removing the prize, those involved in the physical handling of cash are only steps away from removing the crime itself. So, what is this magical new innovation?

A small electronic device is built into the ATM cash cassette which, upon activation, sprays the banknotes nestling inside with a form of cyano-acrylic glue[53]. This is immediately absorbed into the fabric of the notes fusing them together as a solid block. For polymer notes, sulphuric acid achieves the same effect. Attempts to peel off the notes or remove the hardened glue permanently defaces the currency to such an extent that neither a note-accepting machine nor a shop-keeper can avoid noticing a problem.

This glue is spread across the notes in a way that ensures only half the block is fused. This leaves the serial numbers still visible, thus ensuring that the notes, although totally useless to the criminal, can be redeemed via a central bank on recovery. This method, ingenious in its simplicity, not only renders the contents

---

[53] Gluefusion® [http://www.mactwin.com]

of the cash cassette worthless but removes any liability on the part of the cash-in-transit operator. Furthermore, with the glue sprayed only over a limited area, the cassette itself is reusable, with only part of the lid having to be replaced.

The device is built into the lid of the cash cassette and supported by a second device housed within the ATM itself; the two being linked via a Bluetooth® connection. As soon as one of the sensors in either the ATM or the cassette detects an unusual acceleration – as when being tilted, for example, or when being blown up – or detects unusual extremes of temperature, or even a more conventional form of unauthorised access, the small canister of bonding fluid is activated. Once released, the liquid is sprayed along the top of the banknotes, dispensed through a high pressure, purpose-built fluid channel.

All of this makes 'smart' cassettes very popular and user-friendly to security guards and cash handlers as this relatively simple technology has effectively rendered cash unstealable. This is a transformation in itself … the person carrying the cassette across the pavement/sidewalk no longer has to wear a helmet or protective clothing. (S)he can just wave the cassette around yelling, "Hey! Bring it on!" This, in turn, means that armoured trucks no longer need to be armoured and that complicated transfer systems in conflict zones such as Syria – where suitcases of cash are furtively and dangerously smuggled across front-lines – can become a much simpler and more transparent alternative to the more traditional forms of humanitarian aid delivery.

Criminals who attempt to break into 'intelligent' ATM cassettes cannot avoid being sprayed with a mist of SmartWater® either, which cannot be seen by the naked eye. This 'water' contains a specific DNA-style code which stays on the offender's skin and clothing for up to six months no matter how many times they are washed. Such forensic evidence can link a criminal to a specific attack. In the UK, this kind of liquid currently has a 100% conviction rate when used as evidence in a court of law.

Faced with such counter-measures, anyone planning to rob an ATM either in situ or as the cassettes are in transit would want to weigh the risk against the potential reward. In other words, they'd like to know how much cash an ATM cassette holds. The answer depends entirely on how often it is used, the type of neighbourhood in which it's found – and therefore the size of average withdrawal – and the denomination of the banknotes loaded into the cassettes[54].

In theory, an ATM with four cassettes, each capable of holding 2,500 notes, and loaded only with £100 notes can hold £1 million. However, ATMs usually carry a standard mix of £50, £20, £10 and, increasingly, £5 notes and will usually hold about £140,000 when full. Most ATMs, however, hold fewer than four cassettes which, given average usage are half empty at any given moment in time. The ATM you just walked past, therefore, has about £40,000 in it.

It was not the smartphone, but the ATM which introduced most of us to our biometrics for the first time.

Our true self, our identity, is not as tangible and solid as we would like to think. As with a piece of precious metal, our identity becomes malleable and ductile when exposed to different environments. One minute it's coldly solid; the next, it's melting in the heat. One minute we're pretty clear about who we are; the next, we find someone else is trying to be us.

For as long as the ATM has been around, authentication of our identity has been based on things we either possess – such as a key, a passport, or a credit card – or something we know such

---

[54] ATMs in poorer neighbourhoods have lower denomination bills and hold fewer banknotes as they are more liable to physical and skimming attack

as a password, the answer to a pre-determined question, or a personal identification number. Used in combination – a process known as 'multi-factor authentication' – possession and knowledge is generally all that is required to confirm someone's identity. However, as we hear about all too often, these conventions can be compromised as possession of a token or the requisite knowledge by the wrong individual can, and still does, lead to breaches of security.

Historically, accessing our cash from an ATM has always hinged on our ability to remember a four-figure PIN known only to us. But PINs, like passwords, are a pain because they change and we forget them. There are only 10,000 possible combinations that the digits 0-9 can be arranged in to form a four-figure PIN code. And with no additional limitations, it is only a matter of time before the code is cracked by 'brute force'[55].

That said, the chance of someone randomly guessing your PIN after finding your card on the street is extremely low ... but not as low as you might think. Humans are not random number generators, and stupidly use more predictable sequences of numbers than they should when choosing their PINs. Using a real-life database of 3.4 million four-figure PINs which contained every single one of the 10,000 combinations from 0000 to 9999, a statistician recently tried to determine the odds of being able to crack a hypothetical PIN. What he found was quite alarming[56]:

Nearly 11% of the 3.4 million PINs were 1234 with 1111 being the second most popular at 6%, and 0000 at 2%. When he constructed a table of the most commonly found passwords, he found that a staggering 26.83% of them could be guessed just by attempting 20 combinations and there was a one in five chance of

---

[55] A 'brute force attack' is a trial-and-error method used to obtain information such as a user password or personal identification number (PIN). Automated software is used to generate the entire sequence of possible consecutive guesses.
[56] www.datagenetics.com/blog

cracking the code using just the five most popular combinations. Statistically, with 10,000 possible combinations, and passwords which are uniformly and randomly distributed, we would expect these 20 passwords to account for just 0.2% of the total, not the 26.83% actually encountered. One of these combinations was 2580. For a while, he couldn't work out why this would be so popular. Then he realised that 2580 is a straight line down the middle of most keypads. He also found that people preferred even numbers over odd, and that many of the high-frequency sequences started with 19xx … which would indicate that people are prone to using the year they were born, an anniversary, or some other equally important event as their PIN. In fact, every single 19xx combination can be found in the top fifth of the dataset. If hackers were to invert this strategy – and they do – knowing the age of a target would give them a 40-fold advantage when guessing the user's PIN. He also found that a staggering 17.8% of all PINs are repeat-pair couplets, where the second pair of digits replicates the first.

So, in our feeble attempts to remember them, we make PINs memorable, and thus much easier to crack. A 4-digit number was enough to protect our personal data back in the sixties when *Dixon of Dock Green*[57] was enough to deter petty crime, but is no longer secure enough in our hacking, fracking, fraudulent world. Since Einstein himself could not keep track of all these changes without writing them down somewhere, it is clear that we have some way to go when trying to make our on-line transacting secure. There must be a better way.

Guessing PINs is risky for a fraudster, though, especially now that it's not just the sequence of numbers, but aberrant behaviour

---

[57] *Dixon of Dock Green* was a moralistic UK daytime television series aimed at small children where the hero, a mature and sympathetic police constable, a typical 'bobby', thwarted petty crime through common sense and human understanding.

that determines whether the card gets retained by the ATM, or not. Algorithms – a fancy word for pattern recognition in computer coding – can recognise aberrant behaviour either at the ATM console or in the vicinity of the ATM via a camera. A second person's face near the shoulder of the user, for example, will lock the machine.

Secure self-banking that relies on who you are rather than what token or knowledge you possess would certainly be convenient. Such a 'third factor' exists in the form of our physical and emotional selves. This has advantages over both of the others in that the user does not need to remember anything or possess anything physical in order to be identified. Credit and debit cards can be lost or stolen, and passwords and PINs can be forgotten or compromised, but our biometric information cannot be cloned … or can it?

We all have only one true identity and this identity must be protected. The ability to securely link or bind a growing set of digital identities to ourselves with physical or behavioural identifiers that only we possess – otherwise known as biometrics – will not only simplify life but also make it more secure. Intelligently coupling what we have and what we know with who we are is a much better way forward in today's complex digital world.

When all is said and done, however, no single mechanism is foolproof. As will be discussed in more detail later, banknotes have over 35 security features embedded in them, yet only a few are used at any one time to detect fraud. But the genius is that we don't know which few. And, if we did, they are changed from time to time anyway. It's the same with our bodies. Multiples of almost unique characteristics when taken together and at random, confer almost total confidentiality.

There are therefore no longer any valid technical reasons why we should continue to expose ourselves to identity theft through the use of outdated security systems and practices. As a result, and after years of trials, biometric authentication is now proliferating throughout the world's banking infrastructure. The revelation in mid-2016 that the account information of at least 500 million Yahoo users had been stolen two years earlier was a wake-up call that we all need to be smarter about our digital security. This applies to ATMs as much as it does our smartphones, which are increasingly the repositories for all sorts of detailed information about our digital lives.

Unfortunately, we've all come to realise that two-factor authentication is increasingly vulnerable to fraud from card loss, cyber-theft, and skimming. Phil Scarfo, vice president of global marketing for biometrics firm HID Global points out, "… with the growing multitude of digital credentials, identities and passcodes we need to store on cards, tokens and smart devices, we face an even more fundamental problem: None of these digital identities is bound or linked to the actual person. Every new digital identity we create represents yet another detail to manage and, worse yet, a potential threat to our true identity."

"The challenge for those who are charged with protecting our true identity and real assets," he says, "is to strengthen transactional authentication without making ATMs or other systems impossibly hard to use. Biometrics solves the problem of providing the only true means of linking digital identities to 'you' and determining who is actually using the system."

Turning our smartphones into personal data fortresses using ever-longer and more complicated passwords will not be enough, especially as interactions between our mobile wallets and ATMs become more frequent. The fingerprint-based biometric has become the most widely used on smartphones and laptops because of its ease of use, interoperability, ability to thwart casual imposters, and lower cost.

When managed correctly and coupled with intelligent encryption-enabled and tamper-resistant devices, biometric authentication is secure and requires no more than the touch of a finger to assure a financial institution that an authorised accountholder is actually present at their ATM. For many public sector banks, this assurance has an additional benefit: it confirms that the user is still alive. This was very important for Argentina's Banco Supervielle – whose kiosks are used to distribute government pension benefits – as the bank had a significant problem with fraudsters trying to claim their deceased relatives' pensions. To combat the problem, the bank began rolling out fingerprint authentication with multispectral imaging technology in October 2013.

But no single security measure on its own is 100% fraud proof. The concept of layering has therefore become a core principle of any security solution. Combining fingerprint authentication with a second or even third factor provides even greater security, especially at a time when we are incorporating an expanded set of personal devices such as smartphones, wearables, watches and smart cards into a growing identity and access management ecosystem.

In South Korea, more than 26 million customers of BC Card can use the financial service company's mobile payments app to verify transactions by simply speaking into their smartphones. Catering to consumers fed up with fiddly and lengthy authentication processes, including PINs and passwords, FinTech companies are turning to voice recognition to make the transferring of money faster and safer. Visually impaired customers of Abu Dhabi Islamic Bank (ADIB) can use their voice to perform standard ATM transactions and can opt to blank out the screen while doing so for added security. Developments in voice recognition technology are on par with current fingerprint technology, where we have got to the point that the wrong person is identified only once in every 10,000 attempts. Neither, on their own, are completely foolproof, however, and security analysts are unanimous in saying there should be more than one level of authentication – and many layers of defence – before a transaction can be completed.

Typically, smaller transactions – say, those of between $50 and $200 – require only one biometric for verification, normally fingerprint, face, or voice confirmation. Larger transactions require more expensive technology such as palm-vein or iris scanning, often used in conjunction with some form of PIN. But even this is not enough.

It has been suggested that, by adopting multi-factor authentication which use a variety of biometrics such as typing patterns, voice recognition, and palm-vein scanning and blending them with one-time passcodes sent to the customer's mobile device, and, say, current location, the security of payments platforms would be significantly increased. The search is on to do just this and do it in a way that is convenient for a generation of 'digital first' consumers, who use their mobile devices in practically every aspect of their lives, and who rarely, if ever, interact with a live person at a bank branch or online.

This convenience does not come cheap, however, and financial services providers will have to make an economic assessment of how much they will lose to fraud versus how many customers they could lose – or will not gain – due to an over-complicated user experience.

ATM provider Itautec has employed a card-plus-fingerprint approach on tens of thousands of ATMs that support hundreds of millions of transactions per month for one of Latin America's largest private banks. With fingerprint technology, some bank customers even enjoy cardless access for selected, low-value transactions, thereby offering the convenience of making their finger the only required personal key or 'wallet' for withdrawing cash at an ATM.

Poland's co-operative BPS bank was the first in Europe to install a biometric ATM that allowed customers to withdraw cash simply with the touch of a fingertip. Introduced in the Polish capital of Warsaw, the machine incorporated the latest in finger-vein rather than finger-print technology. In the six years since, this technology has become the default biometric option of choice at ATMs around the world, including most of those deployed in China, Russia and Turkey. Interestingly, finger-vein scanning has been deemed to be too expensive for mobile devices, which have instead opted for the cheaper and less robust – and therefore less secure – system of sub-dermal conduction.

"Ultimately," says Scarfo, "Multimodal solutions will combine multiple biometrics with other credentials such as GPS-verified time and place factors. This will enable the kind of seamless authentication that will be attractive not only to the consumer and financial services industry but also to many other vertical markets and government-based applications where identity and security really matter."

Biometrics are revolutionising the way we live in exactly the same way the PIN did fifty years ago. But it might surprise you to know that, outside of the military, it is the ATM, not the smartphone or electronic immigration device, which lies at the cutting edge of introducing us to this technology.

Given that biometric data is used to verify that we are who we say we are, it is becoming clear that a combination of such features is likely to be needed if these characteristics are to be considered absolutely unique. In the near future, it is highly likely that our irises as well as our fingers or palms will be scanned before we can withdraw money from an ATM. Iris recognition technology was being used by Syrian refugees in Jordan's transit camps back in 2012.

The trouble is, each supposedly unique biometric characteristic turns out not to be unique at all, and on their own are not very accurate in identifying us. Apart from our DNA, the only physiological characteristic which is totally unique to us is our heartbeat[58] which is determined by the size and shape of our heart, it's efficiency, and its location in the body.

On the face of it, such biometric data confers greater security, but, handled wrongly, could end up rendering our on-line information far more, not far less, vulnerable. In George Orwell's book 1984, his central character, Winston Smith, is monitored by a giant 'telescreen' broadcasting propaganda messages and recording his movements while every sound he makes is monitored and logged by the dreaded Thought Police. This, of course, raises significant ethical issues about his place in society and the nature of privacy.

With the advent of 'smart' phones which track our movements, 'smart' televisions which record our conversations in the home even when we think they are turned off, and ATMs able to identify us before we've even stood in front of the machine, intrusions into our privacy are becoming the norm. We are all Winston Smiths now.

This is because it's too easy to clone our personal data. Fingerprints can be 'skimmed' – as in the film *Bourne Identity* when Matt Damon's thumbprint is transferred from a wine glass to an optical reader via a length of Scotch tape – and digital versions of our retinas and are easily 'spoofed'. Worse, we fear

---

[58] More accurately, our PQRST pattern; the five peaks and troughs that appear in electrocardiograms (ECGs).

that fingers can be chopped off and eyeballs plucked out to fool biometric scanners into thinking it is us standing in front of them, as was demonstrated by Tom Cruise in the film *Minority Report*. As watchers of forensic crime movies and spy thrillers will confirm, this makes us anxious. Some people, for example, think that the print from a cut-off finger combined with a PIN extracted at the point of a gun will be enough for criminals to access their cash from an ATM. This may well have been the case in the past, but no longer.

For a start, in both cases, biometric scanners can tell if the body part is 'alive'. In the case of our fingers, it is not our two-dimensional 'prints' that matter so much, but the three-dimensional vein patterns beneath them. For finger vein and retinal scanning, blood pressure must be present.

In the case of iris recognition, it gets even more futuristic. The iris contains a unique and complicated pattern of ligaments, fibres, ridges, crypts, rings and corona that, whether expanding or contracting, can be used to identify an individual. These patterns are then encoded through a series of algorithms into a 512 byte 'iris code' using wavelet demodulation to create a phase code similar to that used in DNA sequencing. In addition, not only must there be a pulsing blood supply, but the blood must be oxygenated. This is because the colour of blood changes within seconds of the oxygen supply being cut off and the iris has a different infra-red absorption spectrum when alive and full of oxygenated blood. This can be detected even by a comparatively low-tech scanner. These sorts of scanners can also detect 'proof of life' by checking to see if the pupil is dilating or contracting while the scan is going on. This is because a live iris oscillates at about 0.5 hertz even in stable light conditions and is in almost perpetual motion as it regulates the amount of light entering the eye. And it doesn't stop there.

First, unlike in the movies, it is not one but both irises that are being scanned and matched. Given that no individual's irises are exactly alike – even those of identical twins – the chances of defrauding a machine by presenting two 'dead' eyeballs within seconds of their removal is negligible. And second, it is not just the iris being scanned, but its position relative to the contours of the eyelids and to each other that is also being looked at. This means that, even before you are actually standing in front of the ATM, a combination of facial and iris recognition can be used while you are still some distance away. The matching is completed before you have even begun instructing the machine what you want it to do. Scary stuff, at least to the civil libertarians.

Nevertheless, almost overnight and unseen, biometrics have inserted themselves into our lives. We now need our retinas scanned every time we arrive at Heathrow airport in London, both hands scanned at JFK in New York, and I even had my left thumb scanned when trying to get through the door to the UN office in Sierra Leone. In some cases, our voices are enough to allow telephone access to our bank accounts.

Retinal scanning is often confused with iris recognition but is the only biometric which is more reliable than iris recognition. Although our retinas have only about 40 unique data points, our irises have more than 250. Somewhat counterintuitively, this means that the false positive[59] error rate for retinal scanning is much higher than for iris scanning, at roughly 1:10,000,000 compared to the iris recognition error rate of roughly 1:131,000. But retinal scanning is much more intrusive process than iris recognition, requiring the subject to stay very still for some seconds, with the eye held within inches of the scanner. Iris recognition, on the other hand, does not vary with the distance of

[59] A false positive result incorrectly indicates that a person is authentic. A false negative incorrectly indicates that a person is an imposter.

the eye to the camera, the size of the pupil, whether contact lenses or glasses are being worn or not, or differing camera angles. Bloodshot or jaundiced eyes do not affect it either ... very handy to know when trying to obtain money halfway through a late-night party.

Iris recognition technology does not save an image of the eye as the biometric template, instead it saves a series of data points that serve as a map of an individual's unique iris pattern which can be identified by a proprietary algorithm. It is nearly impossible to reverse engineer an iris template to re-recreate an image of the eye and even if this did happen, the image would be useless with a modern iris camera because of sophisticated 'liveness' detection features that render an image of an eye unusable in the absence of a pulse, oxygenated blood, or muscle movement.

Iris recognition shares many of the characteristics of fingerprinting. Both are reliable and extremely accurate, but iris recognition has a 500 times lower error rate than fingerprinting, mostly because iris data points remain constant throughout a person's life, unlike hands, voices or faces which can change with age, injury, illness ... or by plastic surgery. Fingerprinting also requires you to physically touch a device each time the finger is presented for verification, which some people find not a good idea in a post-Ebola era of disease transmission. In contrast, the iris template is created at a distance without any physical contact.

Fingerprint scanners don't always use vein recognition. Less robust systems – such as those used in smartphones, for example – work by measuring electrical differences between the pores, ridges, and valleys of a single finger or thumb which then builds a two-dimensional picture of the arches, loops, and whorls which constitute your 'fingerprint' by capacitive scanning. Looking at the electrical conductivity between the tops of the ridges and the bottoms of the valleys in the sub-dermal layer of skin located immediately underneath as this is more accurate than looking at the dead surface of the skin alone, which is constantly changing and isn't conductive. Ultrasonic variations can also be used. These prints are different for each digit. Keeping it secret which thumb or finger you choose to use adds another layer of security.

In addition, different recognition systems use slightly different algorithms to identify key fingerprint characteristics, which can vary in speed and accuracy. Typically, these algorithms look for where ridges and lines end, or where a ridge splits in two. Collectively, these and other distinctive features are called minutiae. If a scanned fingerprint matches several of these minutiae then it will be considered a match. Rather than comparing the whole fingerprint each time, comparing minutiae reduces the amount of processing power required to identify each fingerprint, helps avoid errors if the scanned fingerprint is smudged, and also allows the finger to be placed off-centre or be identified with only a partial scan.

In short, fingerprints are not a secure method of electronic authentication and should be used with caution when conducting anything other than small financial transactions. This is not just because of vulnerabilities in the hardware when scanning data points or in the number of false positives, but because of vulnerabilities in the software used to capture and transmit the data. Hackers have already managed to bypass fingerprint security measures on a large scale[60], a situation compounded by the recent cyber-attack on the US Office of Personal Management in which almost a million fingerprints were stolen along with social security numbers and personal details.

---

[60] Black Hat Conference, Las Vegas, August 2016

Fingerprints are a relatively insecure method of electronic authentication for financial transacting.

Natural biometric detection systems measure not just our physical characteristics, but our quirky and idiosyncratic behavioural ones too. How we smell, walk, sign our names, or even type on a keyboard are as much a unique biometric identifier as our voice, hands, faces, or eyes.

All of us are unique in other ways, too, including in the way we spend our money. When patterns in our buying behaviour are looked at over time, this individuality becomes even clearer, which, in turn, makes aberrant – and therefore possibly fraudulent – behaviour stand out. By implementing next-generation big-data summation and aggregation algorithms, ATM operators can now personalise their fraud risk management strategy and ensure the authenticity of their customer.

Behavioural software management systems can detect aberrant behaviours which are outside the norms expected at that particular ATM location. The wearing of a face-mask would fall into this category, for example. Other aberrant behaviours – such as a request to withdraw an unusually high amount, or a sudden increase in frequency of use – are more straightforward to detect.

This blending of the physical and behavioural adds up to the realisation that another parameter is needed. And this parameter is largely about your behaviour, or, more specifically, about the interactions between the multiple behaviours that define your 'character'. The way you walk, for example, is a much more effective way at verifying your identity than a fingerprint or even more secure biometrics such as vein recognition or iris scanning. It's cheaper to do, too.

Putting all these features together requires self-learning algorithms capable of what is called 'adaptive behaviour analysis'. This fraud-fighting engine constructs a profile of the individual using a mass of unstructured 'pathways of information' from which it then identifies behavioural anomalies. Knowing what constitutes anomalous behaviour means learning those 'upstream' behaviour patterns that can be considered normal. This might start with places you visit regularly but then gets steadily more specific by looking at patterns in how you transact, not just in what you buy and the amounts you usually spend but how you interact with the device, be it an ATM, mobile phone, tablet, or wearable. The way you type in your PIN and the way you navigate through the options available creates a virtual footprint that becomes more and more representative of who you are with every touch. The machine remembers. And it learns.

The real benefit of such adaptive profiling systems is that the ATM is able to identify nuanced screen behaviours such as logging in slower or navigating the screen in a slightly different way – the sort of behaviour that might be expected when conducting a transaction under duress at the point of a gun, for example – and take preventive action before funds are transmitted.

Geographic location also matters. Any attempt to access funds from an unusual location – when on holiday on the other side of the world, for example – or from two different locations that could not physically be reached in the time it would take to travel between them, is likely to be blocked. This can still pose challenges for Americans returning East-bound by air to their homeland from Asia, as, having to cross the international date-line, it is possible for them to land before they have taken off. In many cases, transactional software cannot compute this anomaly and any attempt to use the same card to withdraw cash on arrival which was used near the time of departure will trigger an alert … resulting in innocent user's card being retained by the ATM and the LAPD being mobilised.

Carnegie Melon's Biometrics Center has developed technology that can identify a human from 40 feet away just by scanning the person's irises. In addition to adding long-range functionality, the system also only requires a "single off-the shelf" camera. This is instead of the more expensive and bulky multi-camera setups that most commercial iris ID systems use today.

What about the great white hope in the early days of biometric application, facial recognition? Do you remember the story of Dad McAfee, founder of the antivirus software company of the same name, going on the run from the Belize police? According to his blog, McAfee disguised himself by growing a beard, dyeing his hair, darkening his face with shoe polish, and stuffing his cheeks and nostrils with cotton wool. This rather theatrical approach apparently helped him evade detection until he surfaced some weeks later in Guatemala. Could face recognition software have captured him sooner?

Probably not is the answer in reality, facial recognition software is still surprisingly clumsy. Some people do not need to do anything nearly as extreme as McAfee to fool facial recognition systems. In fact, they don't need to do anything at all. That's because certain faces are just too "normal" for facial recognition systems to work with, according to surveillance experts.

To understand why this might be, it's worth considering how most facial recognition systems work. First, they have to work out that they are actually been presented with a face – a process known as face detection – and only then move on to recognition of its unique features before matching it with a face from the database already in the system. Face detection usually involves detecting tell-tale "intensity signatures" of dark and light spots of contrast on an image that resemble the two eyes, nose, mouth, and distances between them of a human face. Computers work in a different way: they don't look for physical features. Instead they look for horizontal and vertical patterns of contrast.

Once a face has been detected there are a number of techniques used to recognise it. One way is to create a mathematical representation – something known as a feature vector – that is constructed from standard faces of different proportions. The feature vector is essentially a recipe for a face that can then be used to match it against others with similar features. These are known as *Eigen faces*. Another major flaw of most facial recognition systems is that they need to use the whole face as part of the recognition process. This means that it's possible to fall the system simply by grinning or pulling a face, a strategy which is extremely unlikely to fall a real person. In addition, putting a scarf over the mouth and nose, or simply wearing dark glasses can fool the system which is why they have to be removed before a scan is initiated.

But what about McAfee's other countermeasures? Neither the skin darkening, facial distortion, nor hair dyeing would have fooled facial recognition software as the intensity signature – the measurement of distance and contrast – would not alter much. But the real weak link in face recognition is what happens to these angles when you're not face-on to the camera. The easiest thing you can do to defeat facial recognition software is to tilt your head up and down or to one side. This doesn't attract much attention but it defeats pretty much any recognition system.

But there is a darker side to biometric verification systems: We are used to trading personal information for convenience. We deal with social media and websites every day. Credit cards require us to give away our data for the convenience of using credit over cash. So far, those trades have worked in our favour, but that's because the systems have always included an escape clause inasmuch that those of us who have been digitally violated can change our social media profiles, alter passwords, or cancel cards. That's not the case with biometrics where the uniqueness of each variable is also its greatest weakness: Once compromised, you can never get your identity back.

Ultimately, it is the consumer that will have the final word in how these technologies will be applied. Trends never take off in the way predicted. The streaming and downloading of digital music was set to destroy the music industry before younger consumers started reverting to vinyl and e-books were set to destroy the publishing industry before sales plateaued in 2015 and readers started buying hard copies again. At the end of the day, it will all come down to how much we as individual consumers will trust the systems by which our identities are verified.

In a survey conducted in Australia in 2012, it became clear that most people were happy to adopt digital habits for most of their banking needs and that they didn't mind intrusive technology if it made their financial lives easier. Just over one third (38%) of those Australians surveyed said they would prefer to live in a world where they didn't need to carry cash; two thirds (66%) said they would be comfortable having their eyes scanned by an ATM in place of having to input a PIN. But there's something in the Australian psyche that makes them comfortable with being 'early adopters'. The same cannot be said for most Europeans. When the survey was published, it stimulated much newspaper commentary along the lines of how "Australians can imagine a world in which there is no longer a need to carry notes or coins, or even plastic cards" and "Australians are comfortable with biometric identification, as it combines convenience with security".

Unfortunately, such sweeping statements are guilty of over-generalising the actual situation, especially when you consider that the people responding would have got most of their understanding of biometrics from science fiction movies. The fact is that technology like this just doesn't work like people think it does. This is arguably more true of biometrics that it is with any other type of automation technology as one of the subtle things about biometrics is that they don't simply combine 'convenience with security' but rather pitch convenience against security. You can't improve one without sacrificing elements of the other. Steve Lockstep, a financial systems consultant, explains it as follows:

"Because of the vagaries of human behaviour and the way this varies from day to day, biometrics have to cope with the same person appearing a little different each time they stand in front of the machine. Vein recognition software gets round most of the inconsistencies – in having temporarily worn out your thumbprint, for example, when recently grouting the bathroom tiles, or by background noise interfering with voice recognition systems – but machines are machines, and sensors get grimy and wear out over time.

So biometric recognition systems cut you some slack; the more slack, the easier they are to use but the sloppier they get. Inevitably this means that a biometric system will occasionally confuse one person with another; that is, it will permit a 'false positive'. So you can quickly see that there is an inherent trade-off in all biometric applications, between their ability to discriminate between different people – which is necessary for security – and the speed – and therefore convenience – with which they do this. We can't have it both ways: a system that is highly secure and discriminating will be more inclined to reject a legitimate user while, conversely, a system that never fails to recognise you will also be more likely to confuse you with somebody else.

And when we imagine biometrics replacing cash and cards one day, we need to be precise about what we have in mind. Do you want to pay for your fuel or your groceries with a simple swipe of your hand? Because if you do, you will do so full in the knowledge that one inevitable day a false positive match will be made and you will be spending somebody else's money ... which might not be so unpleasant until you realise that somebody, somewhere else will be doing it to you, and probably for something of far greater value."

Biometric verification uses not one, but multiple variables, each of which has to be matched against a known database of people who have signed up to the system. The more people who have signed up, the higher the probability of a 'false positive' arising. In Japan, Turkey, and Russia, where palm-vein scanning at ATMs is commonplace, you are also required to enter some numbers, either the last four digits of your mobile phone, a PIN, or the day and month of your birthday. This reduces the chance of things going wrong but does not make the system infallible.

The challenge will be made worse for a more practical reason: Providers of point-of-sale equipment will be under pressure to incorporate engineering trade-offs into the design and manufacture of the equipment used. The cost might be coming down but the hardware and software required for infra-red iris recognition and palm-vein scanning will never be as cheap as for face, voice, or fingerprint recognition ... and compromises will be made. They already are, as the fingerprint scanner in an i-phone 6 is currently demonstrating.

Biometric authentication is not immune to cybercrime, either. Most financial institutions and mobile device operators consider biometrics to be the most secure of all authentication methods available today, and therefore the method of choice. Cybercriminals on the other hand see biometrics as just a new opportunity to steal sensitive information and have already developed skimmers capable of cloning potential victim's physical attributes and devices capable of illegally obtaining data from palm vein and iris recognition systems.

The most effective biometric incorporates some way of ascertaining 'liveness' i.e some way of ensuring that the person seeking to be authenticated is actually alive and not some two-dimensional cardboard cutout. Some of the most obvious involve movement. Facial recognition systems on smartphones, for example, require the user to blink when taking a 'selfie'. But there are already mobile applications available for free that overlay a mask onto a human face. With such an app, fraudsters can easily cut a person's photo from a social media site, lay it over a live face with eyes and lips capable of movement and use it to fool a facial recognition system. Voices can be computer synthesised using early-day artificial intelligence algorithms to the point that it is now possible for someone to masquerade as you over the telephone or when interacting with an ATM. While artificial intelligence systems will undoubtedly make some aspects of life easier, they will also expand the vulnerabilities of the online world, and that includes the ATM world. Cyber-criminals already exploit the best qualities in humans – trust and the willingness to help others – to steal and spy. The ability to fool people online will only make the problem worse.

Cryptography: the art and science of writing (encrypting) and solving (decrypting) codes.

ATMs are networked devices. That is, they are connected by phone line to your host bank account using the Internet. This is how you can use a random ATM to access your funds from half a world away. Because these wires are public, it is not possible to ensure that unauthorised users will not be able to intercept anything transmitted down them, including your bank account details[61]. So, if we can't stop the intercept, then there is only one option available to us, and that is to encrypt the data so that hackers and fraudsters won't be able to decipher that which they have intercepted ... at least, not without using a unique code otherwise known as a key.

Encryption has been used as long as humans have wanted to keep information secret. Before the digital age, the biggest users of cryptography were governments, particularly for clandestine or military purposes. As a computer science, it involves rather more than jumbling up and then reconfiguring the odd random letter.

The Greek historian Plutarch wrote about Spartan generals who sent and received sensitive messages using a scytale, a thin cylinder made out of wood. The general would wrap a piece of parchment around the scytale and write his message along its length. When someone removed the paper from the cylinder, the writing appeared to be a jumble of nonsense. But if the other general receiving the parchment had a scytale of similar size, he could wrap the paper around it and easily read the intended message.

The Greeks were also the first to use cyphers, specific codes that involve substitutions or transpositions of letters and numbers. As long as both generals had the correct cypher, they could de-cypher any message the other sent.

---

[61] When conducting an ATM transaction, the consumer provides the necessary information by means of the card reader and keypad. The ATM forwards this information to the host processor, which routes the transaction request to the cardholder's bank or the institution that issued the card. If the cardholder is requesting cash, the host processor causes an electronic funds transfer to take place from the customer's bank account to the host processor's account. Once the funds are transferred to the host processor's bank account, the processor sends an approval code to the ATM authorising the machine to dispense the cash. Once the cash is withdrawn, the processor then transfers the cardholder's funds into either the issuing bank's account or, if using an independently-owned machine, into the merchant's bank account.

Cyphers are better known today as algorithms, which are the guides for encryption. They provide a way in which to craft a message and give a certain range of possible combinations. Until the mid-1970's, the same key was used both to encrypt and decrypt the data stream. There were duplicate keys in other words. Cryptography geeks call this system symmetric.

The first major symmetric algorithm developed for computers was the Data Encryption Standard (DES), approved for use in the 1970s. The DES used a 56-bit key. Because computers have become faster since the 70's, security experts no longer consider DES secure despite a 56-bit key offering more than 70 quadrillion (70^{15}) combinations. DES has since been replaced by the slower but safer 'Triple-DES' or AES, which uses 128-, 192- or even 256-bit keys. Most people believe that AES will be a sufficient encryption standard for a long time to come as a 128-bit key can have 300^{33} combinations.

But, with technology progressing exponentially, the single algorithm which constituted the key became less and less secure. One key could have been stolen – or the code compromised – without the other key holder knowing. Something more robust was needed which could fulfil the four basic requirements of modern electronic communications security, namely: confidentiality (communications between two or more locations cannot be read during transit); integrity (the communication cannot be modified or corrupted during transit); authenticity (the communication must originate from an identified party); non-repudiation (the recipient must not be able to deny receiving the communication).

That something was developed by a team working in secret at the UK Government Communications Headquarters (GCHQ) and the US National Security Agency (NSA) code-breaking centres in the 1970's and, though devilish in the detail, was conceptually relatively simple: It involved the use of two different keys, not just one. This asymmetric system is known as 'dual key encryption', and is the model used to ensure our data remains secure, not just when using an ATM but when paying for things with a contactless card or smartphone.

Knowing how dual key encryption works should make us feel more secure and confident in our online protection. So, what was this breakthrough? The answer is, 'the three pass protocol'. The best way to understand the concept is to use an analogy. Imagine an estranged couple, Sophie and Tom, who now live in separate cities. Sophie needs to send Tom divorce papers that neither of them want anyone else to know about yet which requires both their signatures. She signs the papers, uses a padlock to seal them in a small wooden box, and sends them via snail-mail to Tom. A few days later Tom receives the box, but doesn't have Sophie's key, so can't open it. Instead, he screws in another clasp and seals the box with his own padlock. The box is now double locked. Tom then sends the box back to Sophie. She then takes off her padlock and returns the box to Tom who then takes off his own padlock, opens the box, signs the forms, replaces them in the box, and repeats the exercise in reverse.

At first read, this protocol seems very time-consuming and heavy on process. But that's because you're imagining padlocks, metal keys, and the joys of the postal service you are familiar with today. Stop for a moment, though, and imagine that the padlocks are combination locks which need numbers, not metal keys, to open them and that the box makes it into Sophie and Tom's hands in milliseconds, not days. And this, more or less, is the system we use – and trust – today.

The critical advantage of Sophie and Tom's asymmetric system is that they have used an entirely public system without ever having to send a copy of their keys to each other; keys which could have been stolen by a corrupt postal worker in transit, perhaps?

As you'd expect, though, the system is a little bit more complicated than described here. For instance, in Sophie and Tom's scheme, the order of encryption (locking) and decryption (unlocking) is not the same. To allow for such differences, commutative cyphers are needed to allow the order of encryption and decryption to be interchangeable, just as when the order of multiplication is interchangeable[62].

Similarly, the possibility of a 'man in the middle' attack in which the communication is intercepted and then modified is not unrealistic. In the postal analogy, Sophie would have to have a way to make sure that the lock on the returned box really belongs to Tom before she removes her lock and sends the box back, otherwise Tom's padlock could have been replaced by a corrupt postal worker pretending to be Tom. To prevent such a possibility, some form of tamper-resistant digital signature is required.

Continuing with the postal analogy, consider a roadside mailbox. It's essentially a locked metal box with an open slot exposed and accessible to the public. Its location is public knowledge. Anyone can go to the mailbox and drop a postcard or letter through the slot. However, only the postman can open the mailbox. If so inclined, he or she could read the postcard but, unless they rip or steam open the envelope, cannot read the letter inside. And even if they do, the letter might itself be written in code. If the envelope is not just glued but sealed with a personal wax seal, any tampering will be visible and quickly noticed.

This might seem a slightly Dickensian analogy, but we use the modern equivalent today every time we enter our PIN into an ATM. The PIN is totally unconnected to the actual transaction that

---

[62] 4x3x2 gives the same answer as 2x4x3

is about to take place; all it does is verify that the person standing in front of the machine is who they say they are. The PIN-pad even has its own separate encryption system and acts, in effect, as the wax seal. It's only when the green 'Enter' button on the PIN-pad is pressed that the electronic 'wooden box' is padlocked and dispatched.

With normal keypads, the PIN entered by the customer is sent in "raw" state via a cable to a separate circuit card module containing encryption integrated circuits. For most countries, this arrangement was satisfactory because the cable and circuit card are located within the secure area of the ATM. In order to decrease PIN theft fraud, however, the card companies now require an encrypted PINpad to be used instead of a normal keypad. This is a sealed module that immediately and locally encrypts the PIN after entry. There are no "raw" personal identification numbers accessible to electronic hackers who might have physically tapped onto wires within the ATM or who might be remotely sensing electromagnetic radiation emitted through the ATM's wiring. Any tampering of the PINpad causes it to permanently disable itself. The unit must then be removed and shipped back to the manufacturer to be reset.

Returning to the padlock analogy, the type of combination padlock available at your typical downtown hardware store is likely to have no more than four rotating tumblers. Assuming each tumbler is numbered 0-9, this means there are only 10,000 possible four-digit permutations ... four digits being what a PIN requires, incidentally. A brute force attack by a criminal intent on unlocking at least one of the padlocks by manually going through every possible permutation might take a few hours, but is not unthinkable. If this were done electronically rather than manually, with each permutation taking one-hundredth of a second, it would only take 1 minute 40 seconds to open the lock.

If the number of rotating bezels were increased, say, to 56, and each bezel had not just numbers but letters, including capital letters, and ten figures, the possible permutations would increase to 72 to the power of 56. To find the correct key would be like finding one grain of sand in the Sahara desert. In other words, it would take too long to unravel to be commercially worthwhile.

chapter five

cash
: ~'ka /ʃ

noun
physical currency (money) in the form of coins or
banknotes, as distinct from cheques, money orders, or
other forms of credit

THERE ARE MORE SLANG SYNONYMS for *cash* than possibly
any other word in the English language. In one shape or form, the
words *brass, wad, wedge, wonga, spondulis, readies, bread, dough,
dosh, loot* and *lolly* all refer generically to cash. The term itself
probably derives from British colonial days in India where local
words such as the Sinhalese for money, *kashi*, often found their
way into everyday usage. Then there are other words which
denote particular forms of money, such as 'bob' for one British
shilling, 'buck' or 'greenback' for a US dollar bill, and 'quid' for a
British pound sterling. Some of the derivations are clearer than
others: The buck, for example, is a hand-me-down from the days
when trade with first nation American Indians was conducted in
deerskins; the 'greenback' refers to early US dollars which were
printed only in green ink on the reverse; and *bread* – including, by

extension, *dough* – was Cockney rhyming slang for 'bread and honey' *i.e* money.

And some have gone on to form colloquial expressions with their own particular meaning. "He's as bent as an nine-bob note," for example, is a straightforward reference to a counterfeit ten-shilling note[63], and refers to someone who is not straight and therefore crooked *i.e* a criminal or 'crook'.

'Quid' has a logic to it, too, when you know that the Latin expression *quid pro quo* means 'to exchange something for something else of perceived equal value'. But some of the others are altogether strange. While 'wad' clearly denotes a packet of banknotes, 'wedge' has no apparent derivation – unless a wad was once used to prop open a door? – and appears to have even less logic to it than, say, 'dosh' or 'lolly'. The word 'loot', on the other hand, is more specific and infers that the money in question was gained illicitly *i.e* was stolen.

Actually, the word 'lolly' stands out as much for its historical antecedents as anything else. These were told to me by an Australian friend, a Quaker, whose uncle had passed the story onto him. And, as his uncle was the man who invented the polymer substrate used in today's plastic banknotes, I have no reason to think that the story is not correct despite the fact that I have yet to meet anybody in cash management or etymological circles who has heard of it.

A couple of centuries ago, at a time when Australia was getting to grips with administering itself as a British colony, a 'tot' of rum was widely used as a unit of currency. This was a sensible move on the part of the authorities as not only would the 'tot' have been a familiar unit of measure to those who had been deported 'down under' on British ships, but any attempt at dilution would

---

[63] The *Ten-Bob Note* was a much loved currency in the UK until the mid-1980's denoting ten shillings, or half a 'quid' (equivalent today to 50p).

have been quickly noticed by a discerning public who would have been quick to assert their own form of 'self-regulation' had they detected any attempt at meddling. Just as with printed banknotes today, the economy, if not social order, depended on the quantity and the quality of its production being carefully controlled.

The only group who the colonial administration felt could be trusted with assuring this across the country was the God-fearing Quakers, a group with a reputation for being even more 'canny' in business than the Scots. This turned out to be an inspired idea, not just because of the Quaker's business acumen, or even their strong 'bush' network of churches, but because the concept of cheating was anathema to the Quaker philosophy. They were also teetotal. Hallelujah!

Inevitably, however, and as is usual wherever control over the production of alcohol is attempted – think the era of prohibition in the US in the 1930's – people find illicit ways of distilling their own. Eventually this 'unintended consequence' had undermined the Australian economy to such an extent that, in the early 1920's, rum production was phased out altogether.

This gave the Quakers a bit of a problem. Not only were they left with the expensive and now redundant means of production, but they were left with a long and equally expensive supply chain which included extensive agricultural holdings in the form of sugar-cane plantations which had been so necessary for the manufacture of molasses from which rum was made.

At this point, Quaker ingenuity quickly came up with a solution, and one which involved only minor adaptation to the existing infrastructure: Pandering to the sweet tooth of the burgeoning Australian public, they transformed almost overnight from the distillation of rum to the mass production of sugar-confectionary-on-a-stick, a form of hard candy which became an instant sensation. For reasons which have been lost in the history of time, the adoring Australian public came up with the colloquial expression 'lollipop' to describe this new form of 'sweet'. Given their common production heritage, it takes only a minor leap of the imagination to see how the abbreviation 'lolly' became a slang term for 'cash'.

Cash is one of the most successful technologies ever invented.

Beyond knowing the different words used to describe money, and before we can understand how an ATM works and what secret marvels of engineering its bland, boxy exterior contains, it would be useful to know a little bit more about the folding stuff it dispenses.

Firstly, what is money? According to James Surowiecki in his book *A Brief History of Money*, economists typically define money in terms of the three roles it plays in an economy:

- As a store of value, meaning that money allows you to defer consumption until a later date.
- As a unit of account, meaning that it allows you to assign a value to different goods without having to make subjective comparisons between them.
- As a medium of exchange, making an easy and efficient way for you, me, and others in-between to trade goods and services with one another.

It's also an abstraction. In the 13th century, the Chinese emperor Kublai Khan – the one in the poem Xanadu penned by a drug-addled Samuel Taylor Coleridge – introduced paper money in an effort to encourage trade within the Empire which at that time was plagued by each region having its own coinage and each coin its own value. The great Khan's daring notion was to make paper money rather than coins the dominant form of currency. And when the great Italian merchant-adventurer Marco Polo visited China at the time, he marvelled at the spectacle of people exchanging their labour and goods from mere pieces of paper. It was, he thought, as if value were being created out of thin air.

Kublai Khan was ahead of his time: He recognised that what matters about money is not what it looks like, or even what it's backed by, but whether people believe in it enough to use it. With metal coins, both the quality and quantity of money in circulation should be fairly easy to control as the raw materials are in relatively short supply and quality can be assayed. When it comes to paper, however, an altogether different situation applies as why should we believe that a piece of paper is actually worth the number printed on it?

Like Peter Pan asking us to save Tinkerbell by joining him in believing in fairies, the ability to convince other people to believe in something they can't see but wish to be true, requires a fairly substantial act of faith. But we do it all the time. In fact, as Neil MacGregor, director of the British Museum in London, points out in one of his fabulous BBC radio essays on *A History of the World in 100 Objects*, the banking system of cash and credit we use today is built on this simple act of faith. As much for this reason as the technology used when printing banknotes, he says, "Banknotes are truly one of the revolutionary inventions of human history."

Since we all want something universally recognisable as money – we don't want go around having to judge whether this type is worth more than that type – it makes sense for the state to become the only issuer and guarantor of paper money. For this to work in principle, counterfeiting has to be discouraged in practice because an enormous gulf exists between the low 'real' value of the piece of paper and the 'promise' value printed on it. This is done in two ways: by making the production process so difficult that it becomes impossible to forge, and by making it illegal to attempt to do so. The Ming banknotes of Kublai Khan incorporated words like "to counterfeit is death" into their designs and anyone outing a criminal forger was handsomely rewarded. Production of the specialised papers and inks involved were carefully controlled, too, as they are today. In fact, banknotes are not printed on paper at all; Khan's banknotes were printed on mulberry bark and today's banknotes are made from either plastic or a mix of cotton and linen[64].

Imagine what trade would be like in the absence of the folding stuff? We'd still be herding sheep across London Bridge or down Pennsylvania Avenue to barter in the market for sacks of wheat, which we'd then have to carry all the way back home again. The chief problem with this form of transacting is what economist William Stanley Jevons called the "double coincidence of wants." Say you have a bunch of bananas and would like a pair of shoes; it's not enough to find someone who has some shoes or someone who wants some bananas. To make the trade, you need to find someone who has shoes he's willing to trade and who wants bananas. Unless you happen to be wandering shoeless through a stony banana plantation, that's a tough ask.

With a common currency, though, the task becomes easy: you just sell your bananas to someone in exchange for money, with which you then buy shoes from someone else, maybe somewhere and sometime else. If conducting such a trade with a foreigner, having a common medium of exchange would be even more valuable. That is, money is especially useful when dealing with people you don't know and may never see again. But only if you can both count, and only if you both agree on its relative value. Because we do, money is one of the constants in human affairs. From the gold coins of Croesus and banknotes of the Ming

---

[64] In homage to this little piece of history, a Mulberry tree grows still in the courtyard at The Bank of England.

Dynasty to the virtual currencies circulating digitally today, money is nothing more than an expression of how far a stranger can be trusted. Money was not just invented as evidence of a promise to pay later, but to get around the problems of trusting people you might not know by instilling confidence that you will be paid. As Mervyn King, the former governor of the Bank of England put it, "money is essentially an IOU where the 'U' has to trust the 'I'."

Currency has taken a number of forms in our time, one of which, salt, is still used in China and Somalia.

Our past is full of revolts and wars over the manufacture and distribution of salt. An unpopular salt tax was a significant cause of the French Revolution, for example, while Napoleon's defeat in Russia was blamed at the time on the lack of salt. The British were credited with nearly turning America's War of Independence when they succeeded in capturing George Washington's salt supply.

One of the pivotal reasons for salt playing such a role in the history of nations is that, for millennia, salt has been used as a form of currency where its quality and quantity had to be carefully controlled by those wanting to remain in power. In ancient Greece, for example, salt could be exchanged for slaves – a bartering process which gave rise to the expression, "not worth his salt" – but also because, in areas where it's not easily obtained, control of its supply confers political advantage, if not untold wealth.

If full production is concentrated in a few places within a country, or if it has to be imported, then it is always possible for its distribution to be controlled. Once a monopoly is created, it is easy for those in control to impose a tax. And history is full of stories about the levying of such taxes, not just during the French revolution but by the British during the colonial days of the Raj in India, and to this day in China.

In fact, salt production in China is perhaps the world's oldest monopolised industry. It is still illegal for any entity except the China National Salt Industry Corporation to sell salt for household use in mainland China. This situation has operated more or less continuously for 2,132 years since it originated in 119 BC during the reign of Emperor Wu Han. Today, over 25,000 special *Salt Police* wearing distinctive red epaulettes and gold badges in the shape of salt crystals do nothing other than enforce this monopoly – one of the oldest economic policies in the world – by monitoring its local production and national distribution. The situation was not so different in Japan where the Japan Tobacco and Salt Corporation – tobacco was also subject to state control – remained a complete state monopoly under direct Ministry of Finance authority until as recently as 1985.

As early as 2,200 BC, taxes were being paid in China in the form of salt. By 900 AD the salt tax formed the largest component of government revenue. Salt was observed by Marco Polo being made up into small cakes way back in the 13th century in Tibet, with "the quality of each cake being carefully controlled by specially selected offices before being stamped with an impression of the Grand Khan". These cakes were, in effect, little different to metal coins.

Aristotle believed that primitive barter trading of standardised commodities, 'hall-marked' by an authority for correct weight and quality, represented the first use of "money". Key to this concept was the term 'hall-marked', being some form of marking to guarantee the quality and quantity of salt in terms of its perceived value.

It is in Europe, however, that the closest parallels with the salt tax in China and, later, in India can be found: the Roman state took over salt production at the mouth of the River Tiber in 506 BC. Roman soldiers received an allowance to purchase salt … the *salarium*, from where the modern word 'salary' is derived.

Salt was also taxed in India from time immemorial. We know, for example that King Chandragupta, who ruled much of India from 324 to 301 BC, imposed taxes on salt. His treatise on governance, the *Arthasastra*, goes into incredible detail in listing the duties of the state's officials, including specifying the times of day that state elephants are to be bathed and what type of women should be recruited as spies. An entire department, headed by the court's most senior official, was responsible for salt. Licences for manufacture were issued for a hefty fee, or in exchange for one sixth of the output. Taxes were also imposed on imported salt.

By the mid-19th century a commercial agent of the British government, the East India Company, administered India and collected taxes on its behalf, the two main ones of which were those levied on land and on salt. Nothing much had changed in over two thousand years, and *The Company*, as it was called, went to extreme lengths, as Chandragupta had done, to enforce their monopoly and thereby control prices. In such situations, others will always try to supply a commodity more cheaply … and this, of course, was duly attempted.

There were three main ways to obtain salt illegally: it could be stolen from government warehouses; it could be surreptitiously manufactured, either at inland salt works or along the coast; or it could be smuggled in from outside. This last option was to prove the greatest threat to The Company's revenues. India's inhabitants, especially in greater Bengal, were desperate for affordable salt yet The Company was determined to raise the maximum revenue possible and block off any illicit supply. The stage was set for an epic battle, which led to the creation of what was surely one of the greatest *folies de grandeur* in imperial history, the creation of an impenetrable thorn hedge which ran for over 2,500 miles from the Himalayas in Nepal to the coast on the Bay of Bengal.

Manned by over 14,000 men deployed in 1,727 guard-posts, the hedge comprised a wall of prickly pear, thorn, and cactus 4 metres (12 ft) wide at the bottom and nowhere less than 3 metres (10 ft) in height. By 1850 it had become one of the greatest constructions of all time, rivalling the Great Wall of China in length. Incredibly, very little is known about this 'customs hedge' except that it's officials harassed the population and extorted bribes, thereby acting as a continual reminder of what Indians saw as unjust British taxes; a situation which Mahatma Gandhi would take advantage of less than one hundred years later in 1930 … as my Dad was witness to himself.

Serial numbers appear twice on banknotes
courtesy of King Henry V of England.

By introducing the 'tally' system – the word being derived from the Latin *tessera* meaning 'treasure' – the next essential element in the evolution of money as currency became possible. Tally sticks were used by King Henry V of England in the 15th Century to pay the 6,000 (mostly Welsh) longbow-men he took to France in 1414, and who were so instrumental in defeating the French at the *Battle of Agincourt* less than a year later. Short of funds with which to pay for his military adventure across the Channel, he had to find some way of 'promising to pay' his army before it marched. His solution was ingenious.

On signing up, each soldier arrived with a foot-long sliver of dried willow or oak which then had his name carved into it and notches cut at various intervals according to his daily rate of pay. This was the tally stick. These were then split down the middle, with the smaller piece, the foil, being returned to the soldier while the larger piece, the stock or counterfoil, was secured in the Treasury at the Tower of London. This, incidentally, is where the financial terms 'stock' and '(counter) foil' we still use today originated. It also describes the situation of the hapless soldier who was expected to risk his life on nothing more than a King's 'promise to pay' after the fighting was over; a disadvantageous scenario known as "holding the short end of the stick."

When the soldier returned, the two matching halves were reunited and he was paid what he was owed. With each tally being unique because of the way the wood was knotted, an individual guarantee was possible.

This was a clever move, as not only did it give Henry credit without recourse to money-lenders who charged usurious rates of interest, but it kept his soldiers loyal. It also earned him additional income as he knew they wouldn't all return. In the event, after arguably the most successful battle in British history, all bar 320 did, but by then he had amassed enough ransom and spoils of war to pay them in full.

As a further refinement, parchment records of the tallys' existence themselves became tangible and transferable to the point that other forms of work and services could be "notched up" as transferable value. Over the next four hundred years, paper 'value' began to replace the wooden tally system, with tally sticks themselves only abolished in 1826 [65] . As banknotes could intentionally be ripped in half as an outstanding 'promise to pay' – a practice which continued in Scotland until the mid-1960's – unique serial numbers were written twice onto each note. This practice continues to this day even though it is no longer legal to deface a banknote by intentionally ripping it in half.

---

[65] The burning of these old tally sticks resulted in the accidental destruction of the British Houses of Parliament in 1834.

Deposits of stored value goods such as salt, hard cheeses, champagne, vintage port, and smoked hams, were also represented by the tally, which then itself became a means of payment. To this day, some of these stored value goods are held by banks, not producers or merchants, as collateral for loans. Genuine Parmigiano cheeses, for instance, are owned by, and mature in warehouses owned by, three banks in Regiano, a province in Italy.

The tally was thus treated as currency. Metal coinage and paper money both became mass produced tallies, cheap to make but also easy to imitate, and so eventually limited to "smaller money" and used to pay individual tax.

With this simple concept, value became an idea, something detached from the intrinsic nature of the thing itself. It could be calculated for different categories of goods, and, more than that, it could be written down, arithmetically juggled, and turned into ratios and equations. A new way of thinking was born, transactional and every day, and yet with momentous implications for the way we live now. Put another way, the reinvention of money a few hundred years ago played a key role in the development of abstract, scientific, and eventually, secular thought. No wonder Neil MacGregor at the British Museum sees money as one of the world's greatest ever inventions.

Money is a collective act of the imagination.

The first paper money in what is now the United States was issued by the Massachusetts Bay Colony in 1690. It was valued in British pounds. It wasn't until the 1760's that the first dollar bills were issued. During the American Revolution, the fledgling Continental Congress issued Continental Currency to finance the war, but widespread counterfeiting by the British and general uncertainty as to the outcome of the revolution led to massive devaluation of the new paper money.

Stung by this failure, the US government did not issue paper money again until the mid-1800's. In the interim, numerous banks, utilities, merchants, and even individuals issued their own bank notes and paper currency. By the outbreak of the Civil War there were as many as 1,600 different kinds of paper money in circulation in the United States, as much as one third of it counterfeit or otherwise worthless. Realising the need for a universal and stable currency, the United States Congress authorised the issue of paper money in 1861. In 1865, President Lincoln established the Secret Service, whose principal task was to track down and arrest counterfeiters.

Similar things were happening in Europe. In 1694, for example, the Bank of England was established to raise money for King William III's war against France. Almost immediately, the bank started issuing notes in return for deposits. What made this a viable means of raising funds was the King's "promise" to redeem the note for its equivalent value in gold whenever the "bearer" made such a demand. It was, in other words, a form of guarantee. But this 'guarantee' was always illusory, and as soon as it was widely understood some 50 years ago that convertibility into gold was merely a method of controlling the amount of a currency, which was the real factor determining its value, governments became only too anxious to escape that discipline, and money became more than ever before the plaything of politics. For it to remain that way, it was crucial that government controlled the printing presses, and exacted severe penalties on anyone who attempted to print banknotes without their permission.

At its most surreal and existential then, money – the folding stuff we hold in our hands – represents a collective act of our imagination. It works because we believe that it works. It has value because we believe it has value. When the value of currency was pegged to the Gold Standard and banknotes could be redeemed for finite quantities of gold, this wasn't the case: A banknote had 'value' because the gold existed and had value. With the abolition of the Gold Standard in 1931, our money became fiat money – the word 'fiat' being derived from the Latin for "I believe" – and it's this that we carry around with us today. This doesn't mean it's worth the weight of a small and under-powered Italian car of the same name, but means we have committed to an act of faith, an illusion, and are confident that 'real' paper money will retain the notional value it had when we first put it in our bank account and that our government will somehow 'guarantee' its value. This, as any Greek living through the financial crisis of 2015 or Serb living through the Balkan Wars of the 1990's will tell you, cannot be relied upon.

Nevertheless, banknotes represent a debt, a 'promise to pay'. You are the 'bearer' and therefore the owner of every banknote you hold in your hand and is for you to redeem for whatever purpose you and the person with whom you are negotiating can agree between yourselves. This is why, printed somewhere near the signature, most banknotes still contain the phrase, "I promise to pay the bearer on demand …". This quaint phrase is not a mere historical echo from King William III's day; the reason the language is there is to signal to all parties in a transaction that the government has taken on the responsibility to ensure that the purchasing power of each banknote is maintained. It is this that builds the confidence and trust in the value of an otherwise worthless piece of paper.

*Modern banknotes incorporate over 35
different security devices.*

For the public to trust and want to continue to use cash, genuine banknotes need to work first time, every time. Cash that has to be tried several times in a vending or ticket machine to be successful would undermine our confidence in the money supply and would make us suspicious that all forms of money had somehow been compromised ... and with it, perhaps, the economy itself. It is therefore important that paper money cannot be printed by anybody who just happens to be "short of a bob or two", otherwise our financial system would have to revert to cowrie shells and gold coins ... which are difficult to dispense through an ATM. This means that some form of secure printing is required if counterfeiters are to be thwarted.

Several well-known early Americans applied their printing skills to outwitting counterfeiters. Benjamin Franklin, a printer by trade, came up with the idea of using tree leaves to print a vignette on the back of banknotes, based on the observation that no two leaves have exactly the same pattern of veins, thereby making the duplication of a banknote impossible. Impressions used were entered into a book, which could then be consulted for comparison whenever a doubtful banknote was encountered. Such forms of innovation – albeit a bit more high-tech – continue to this day.

As an anonymous medium of exchange and as a valuable negotiable document, banknotes must contain – and be seen to contain – exceptional security features if the confidence and trust of the general public in them is to be maintained. Accordingly, a wide range of devices are incorporated into five distinct areas of a banknote: the paper, its design, the way it is printed, the incorporation of 'optically variable devices', and the integration of machine-readable features.

In addition to the many inspections that occur during the printing process, the raw materials are also subject to strict inspections before they are used. The inks are subject to rigorous quality control checks for colour and viscosity, for example, while the paper is tested for chemical composition, thickness, and other properties such as flax content. The paper is produced by a licensed manufacturer in a secret, tightly controlled process. It is illegal for anyone else to manufacture or possess this specific paper.

The design process has over 65 steps, from engraving of the multiple printing die – itself a process involving at least three engravers, none of whom ever meet each other – to the positioning of security devices, with no single person knowing what is being done during each of the steps. An engraver might be told, for example, to leave a gap of a certain size in a certain place, but s/he won't know what device is to be included there, and, while the 'master' engraver knows what device will end up in that position, (s)he won't know its technical aspects.

Every banknote currently in circulation around the world incorporates up to 35 different security features, each of which can be employed in any number of locations on the note, and in any combination. They are scanned regularly to detect counterfeits. Lower denominations last approximately 18 months in circulation and will probably be scanned four or five times in that period. Higher denominations, such as the $100 bill or €50 note, are handled much less often and therefore last in circulation for years. These may be scanned every four years or so.

Because next-generation ATMs now have the capacity to accept banknote deposits and then recycle the same notes to other consumers, they must ensure that they do not become the route through which forgeries begin to circulate. This presents two enormous technical challenges for the cash management industry:

The first is that security printers acting under license of the relevant note-issuing authority – the Bank of England in the UK, the European Central Bank in the case of European Union member states, and the Federal Reserve in the US – must ensure that every reasonable effort is made to ensure that banknotes cannot be forged; the second being that the ATM industry has to adapt the ATM's mechanics to ensure that any such forgeries can be detected.

By code-of-conduct or by law[66], deposit-taking requires the authentication and validation of the notes ... often of different currencies and in multiple denominations. A multi-currency recycler which automates the acceptance, authentication and validation of banknotes while making them instantly available for dispensing to the next customer is complex and expensive technology which is only cost-effective at high volume locations where international customers are crossing borders or most likely to be in need of foreign currency.

First, the notes being deposited have to be sorted. This is technologically challenging as they might have been inserted in multiple currencies and multiple denominations, back to front, upside down, and either vertically or horizontally. An increasing number of recycling ATMs have the capacity to automatically orient the note correctly.

Larger cash-recycling ATMs – i.e those that handle above-average transactions on a daily basis, or are embedded in the walls of a bank branch – can have up to 30 kinds of sensors installed which will be looking out for any combination of the following security features:

---

[66] For example: Article 6 of Council Regulation (EC) No 1338/2001 for the protection of the euro against counterfeiting

Paper: That particular 'crisp' feel of money comes from at least three different things that make the paper in banknotes unique. Normal paper that you use on a day-to-day basis is made from the cellulose found in trees. Paper used for money, however, is made from a special blend of 75% cotton and 25% linen. Security features are incorporated into the paper while it is being manufactured, making them part of the substrate itself. It is also thinner than normal paper and has a very specific colour when looked at under ultra-violet light: most banknotes appear dark blue, sometimes with flecks of pink, while ordinary paper shines bright blue.

In January 1988, Australia printed the first banknote in the world to be produced on polymer substrate. It took AU$20 million and 20 years to develop. Although primarily developed to combat the risk of counterfeiting – polymer banknotes allow for incorporation of many security features that can't be applied to paper banknotes – there are other major benefits to using polymer over paper: The notes stay cleaner and last more than four times longer than paper. You can accidentally put them through the washing machine and all that happens is that they come out cleaner without even a trace of a crease. Being cleaner, they are less likely to carry disease and deposit less muck and ink on the sensitive sensors inside ATMs which means there are less machine jams and therefore lower downtime. This means that Australians have even more opportunities to get hold of their 'little plastic drinking vouchers'. Much more important, though, for some Australians is that their money is still usable after falling into the swimming pool.

Polymer notes are more secure than the traditional kind as not only is the plastic able to hold bumps and ripples throughout the life of the note, but they incorporate clear windows into which are embedded diffraction gratings. When you hold a note up to light, these gratings split and diffract light into different coloured beams in exactly the same way as sunlight is distracted by water particles in the atmosphere to create a rainbow. However, polymer banknotes have downsides as well. They cannot be easily folded, have an unnatural feel – a slipperiness which sometimes makes them difficult for an ATM to count properly – and can be permanently damaged if exposed to the heat of an iron or tumble dryer.

Embedded Security Fibres: Tiny synthetic or treated silk fibres are dispersed throughout the paper during manufacture. They are available in a range of colours that are visible in normal light and fluoresce under UV light. Some of these fibres conform to a wire mesh pattern during the forming process when the paper is still wet

Intaglio Printing: Banknotes are printed using a process called 'intaglio'. This is just a fancy word meaning 'raised'; an effect which becomes possible when up to 60 layers of ink are printed at the same place on the note. Each layer is added with over 60 tonnes of pressure – more than ten times the pressure used to print this book – thus acquiring volume which you can feel with your fingers.

Security Threads: Security threads are among the most common and reliable features, as they are relatively cheap to produce but extremely difficult to counterfeit. There are two common types: internal and diving. Internal threads are embedded within the layers that make up the note. They may be solid or transparent and can be made of metal or plastic.

The diving threads are usually made of foil. These are the ones you can see glittering on the face sides of the banknotes. Usually these threads have micro-text in them – which can be seen, but not read without a magnifying glass, when held up to the light – and holograms. These threads are 'machine readable' which means an ATM can distinguish the currency and denomination from this one feature alone.

Where these threads are located on the note, their width, and the images imprinted into the strip are unique to each currency and denomination. These strips glow different colours under ultra-violet light. With US bills, for example, $5 bills glow blue, $10 bills glow orange, $20 bills glow green, $50 bills glow yellow, and $100 bills glow red. The absence, colour or intensity of this glow helps to determine whether the bill is counterfeit or not.

Using a unique thread position for each denomination guards against certain counterfeit techniques, such as bleaching ink off a lower denomination and using the paper to 'reprint' the note at a higher value.

Some threads include an 'issuing authority legend' which can be read under certain wavelengths of light. Others are coated with thermo-chromatic pigment that changes colour when heated, for example by rubbing with a finger.

Watermarks: These are incorporated into the paper during manufacture and are nowadays three dimensional. There are two common types of watermarks: localised – typically a portrait – and whole area marks, which cover the entire banknote. Portraits are considered more reliable as a security feature as they contain more semi-shadows. The standard watermark is formed by varying paper density and thickness in a defined area during the papermaking process. The image is visible as darker and lighter areas when held up to the light.

Each banknote may have three or more such watermarks, with each occurring at different levels within the paper. Some overlap. There are also digital watermarks – such as the EURion constellation in higher denomination Euro notes – where a digital watermark is embedded in the banknotes' design as part of the printing process.

Watermarks can also be specifically designed to accept magnetic ink character recognition (MICR), and if genuine cannot be seen under UV light.

Portraits: Most banknotes include portraits of the country's ruler and/or someone famous. Such portraits are a brain boggling, eye popping patchwork of fine lines and tiny dots nearly impossible to accurately copy. The larger the portrait, the more detail can be incorporated, making it easier to recognise and more difficult to counterfeit. The portrait is usually shifted off-center to provide room for a watermark and unique 'lanes' for the security thread in each denomination. This also reduces wear on most of the portrait by removing it from the center, which is frequently folded.

Low-Vision Device: This is the large dark numeral on a light background which, by being large and high contrast, helps people with poor vision by being easier to read.

Barcode: These allow scanning devices to identify the note's denomination. Barcodes are quite distinctive and can usually be seen near the watermark when the note is held up to the light. The number and width of these bars indicates the value of the note. Some newer notes use square codes.

Microtext: This print appears as a thin line to the naked eye, but the lettering, being between 0.15-0.18 mm (0.006-0.007 inches) high, can easily be read when using a magnifying glass. The resolution of most current copiers is not sufficient to copy such fine print. Such microtext can be hidden anywhere on the banknote, in the ornaments, around the portraits, in wide strips or single lines, or just merged into the background. To make things even more difficult, the letters can be of different colours and ink densities.

Microporosity: If you hold a Swiss Franc banknote up to the light, you will be able to see daylight through the note ... one of the denomination numbers is not printed, but made up of tiny holes. More sophisticated banknotes, particularly those made of polymer, incorporate patterns of much smaller laser drilled micro-holes which cannot be copied because they cannot be seen without a particular type of laser scanner. Not only is each hole so small that water cannot pass through it, but each is a slightly different shape and size.

Matching Elements: An extremely difficult thing to counterfeit is the matching element. This is when two different parts of the same figure are printed on separate sides of the banknote, which only merge into one whole symbol when held up to the light. It's nearly impossible to get the two elements to match without use of highly sophisticated security printing presses of the sort that only licensed companies such as De La Rue are allowed by law to own.

Background Decoration: Background decorations and ornaments usually consist of several layers, with each layer printed over the one below. The ornament on the layer is usually very small and sometimes simple, but when several layers are combined, usually in different colours, they form a very complicated background.

Gradient Colouring: Gradient colours are nearly impossible to reproduce even with the best copying equipment. The colours blend and grade so gently that the finest photocopier will only be able to discern strips of different colours.

Latent Image: Incorporation of a latent image is a relatively new feature, usually only added to higher denomination notes and made with the help of intaglio printing. Usually the image is hidden in the areas of the same tone and is 'built' of raised parallel lines, which run at an angle to the raised lines of the background. When you rotate such an area with the light reflecting from it, at certain angles you will see the picture appearing and then disappearing.

Holograms and Kinegrams: Introduced in the late 'seventies, holograms and kinegrams are still one of the most advanced security features of today. Holograms show a volumetric, quasi-three-dimensional image, while kinegrams change colour depending on the angle from which you are viewing them. Because they can only be produced by extremely expensive and rare presses, each device has a feature incorporated which makes it possible to tell which press printed which device. Some notes have an additional feature in the form of a silver foil containing holographic images. The images and colours change depending on the angle of observation.

Serial Numbers: Although first introduced to ensure that every single note printed could be accounted for – which is still its primary purpose – and that its source could be identified, serial numbers are themselves security devices. Each banknote's serial number is unique, and they never repeat. There are many different formats, with newer notes using numbers and letters of different colours running vertically or horizontally, sometimes in ascending sizes. Some countries use a mixture of letters and numbers, and some use just numbers. Apart from periodically altering the font, size, colour and typography, these numbers can have mathematical security features built into them as well.

Serial numbers are almost always duplicated on the left and the right-hand-side of one face of the note. As explained earlier, the reasons for this are largely historical even though banks will tell you that it's to enable them to verify that a partially destroyed note can be replaced. That said, it is interesting to note that there is only one serial number on the new range of Australian dollar bills, which may indicate that any security associated with its repetition is now becoming obsolete. Euro polymer notes repeat only the last six digits.

The letter and number sequences tell those who want to know where the note was printed, not just in which country but on what press and using what plate. The numbers also define the row and column location of each and every banknote on the individual sheet of paper on which it was printed.

Being unique, serial numbers from batches of stolen notes are known and can therefore be 'blacklisted'. In much the same way that car-park barriers can now recognise your car's number plate, a scanner will recognise such numbers if they are deposited into a cash-recycling ATM and will automatically alert the police to any such attempt to do so.

Inks: The four denominations of UK money take 85 inks to print. Some of these inks have metals suspended in them which give off a unique metallic glitter as compared to the matt effect seen with other inks. Some fluoresce (glow) when seen in ultra-violet light – go to any European disco, flash a Fifty Euro note and you'll see pictures you don't see under normal lighting conditions – and some do the same under infra-red light.

For US Dollars, black ink is used to print the front of the bills, and green ink is used on the backs ... thus giving rise to the term "greenbacks".

Colour-shifting inks change colour when the note is viewed from different angles. On US dollar bills, the ink appears green when viewed directly and changes to black when the note is tilted. On Euro notes, the colour will change from purple to olive green or brown.

Metameric inks work on the principle where two colours matching under one set of lighting conditions can appear quite different under another. Under normal viewing conditions, nothing is apparent but when viewed under a red filter a numeral appears.

Tiny magnetic capsules are embedded in some of the inks used to print high-denomination banknotes. This magnetism can be detected by fraud authentication devices in ATMs as well as by ticket machines on the New York subway. The capsules also provide a visual cue when coming into contact with a magnetic field, even a weak one such as is produced by magnets in the loudspeakers of mobile phones.

When a bank note printed with magnetic ink is placed over a mobile phone – or any magnet for that matter – the printed image turns clear as the capsules tilt like window shades to let the light through. This is not such sophisticated technology, and taps the same micro-encapsulation methods used by 'scratch 'n sniff' perfume samples in glossy magazines.

Optically variable inks (OVI) contain tiny flakes of special film which changes colour as the viewing angle is varied. They are very expensive and are generally only used in small areas.

Thermochromatic ink either disappears or changes colour when above 88°F (31°C)

Colour Changing: In addition to inks which change colour depending on how you look at them, there are other ways of achieving the same effect. One is to emboss the note with tiny, raised polygons and then print each side with a different ink. When you change the angle, you stop seeing one colour, and see another.

Embossing: Marks for the blind are closely involved with intaglio printing as the raised ink is what blind people can feel with their fingertips. Depending on the denomination, the "finger picture" will differ. Canadian banknotes have magnetic marks (viewed as black marks on the reverse) and when installed in a special reading machine, a voice tells the denomination. The edges of Australian notes are roughened in certain places for the same reason. And most banknotes printed after 2005 now incorporate embossed 'Braille' marks.

Paper Toning: Paper toning is one of the most widely spread security features. The idea is to make the banknote paper have its own unique colour tone. And the more delicate the tone, the more difficult it is to reproduce. The additional bonus of such a feature is that it's easier to distinguish the banknotes of different values by colour. The three Scottish banks authorised to print their own money, for example, have all agreed to issue notes of specific value in the same predominant colour. Thus £5 notes are blue, £10 notes are brown, £20 notes are maroon/purple, £50 notes are green and £100 notes are red.

Matted Surface: A small part of the note incorporates a vertical band which is only visible when lit at an angle of 45°. This tends to be used for the lower denomination notes.

Size: Most banknotes – but not US dollars which, somewhat confusingly, are not only the same colour but the same size – increase in size as the value of the denomination increases.

Tactile Edges: For example, the €200 note has vertical lines running from the bottom centre to the right-hand corner, and the €500 note has diagonal lines running down the right-hand edge.

Chemical Sensitisation: During manufacture, an invisible dye is included which produces a vivid stain when chemicals commonly used to tamper with documents are used by counterfeiters.

See-through Window: Each family of polymer notes has its own see through feature. When held up to the light, the clear areas of the features fill in with colours.

Dot Matrix: Security features in the paper and or in the ink leaves predefined dots of different sizes and of different spacing. These can be clearly seen on a €50 note under UV light.

There are three levels at which banknote authenticity is verified. The most obvious (overt) includes features visible to the naked eye like the watermark and shape-shifting holograms; the second level includes features which need special equipment such as a UV lamp or iodine pen[67] to authenticate them (covert); and the most discreet (forensic) level includes features known only to a very few people, and which require testing by specialists using specialised machinery.

Verifying and identifying counterfeit banknotes at the covert and forensic levels involves sophisticated three-dimensional authentication technology using varying combinations of scanning devices. Visible, infrared, and ultraviolet light – all of different frequencies and wavelengths, and some by laser – as well as magnetic imaging and radioactive detection micro-technology are used for this. Apart from scanning the colour, dimension, composition, serial number and thickness (wear and tear) of the paper, the more than 35 security features embedded, printed or embossed into or onto each banknote is subjected to a

---

[67] This is the pen sometimes used by merchants to verify higher-denomination notes. Iodine turns brown when it comes into contact with lignin in the paper. Since banknotes are not made with standard lignin-based paper, a brown mark indicates a counterfeit note.

varying number of match tests by a cash-recycling ATM. These units can process at least eight notes per second of up to 120 denominations of 70 currencies.

Since the parameters are different for each currency and denomination, detectors must be pre-set for each item the machine is programmed to accept. Several different techniques are available and usually more than five are used. But the settings and sensitivity of each can be changed. The most common form of currency validation technology involves the following scanning techniques:

Optical: ATMs contain small light-sensitive, photo-voltaic sensors as well as miniature digital cameras. These optical sensors look at any number of the security features just described and determine whether the cash being inserted conforms to pre-set templates and is therefore genuine. Certain inks are also fluorescent and glow when ultra-violet light is shined on them.

Magnetic: Cash is magnetic and this magnetism can be detected by what are called 'Giant Magneto-Resistance' (GMR) proximity detection devices. Inserted notes are passed over an array of small magnets which can detect not just the presence, but the orientation of the iron particles in the ink. This results in a recognisable magnetic signature. Hold a banknote near a strong enough magnet and you might be surprised to see that it is attracted.

Radioactive: Some higher denomination notes use radioactive inks. The tiny amounts of radiation emitted – thousands of times less than that to which you are exposed when flying, for example – can be detected by small Geiger-Muller counters.

Size: The thickness, length and breadth of banknotes are different for each type and denomination. Notes are typically 0.0042 inches thick and weigh slightly less than a gram. British £5 notes weigh 0.812 grams when printed but are nearer 1 gram in weight when dirty. The thickness is measured by laser scanner or by measuring the gap between two rollers as the note is inserted. Coins are evaluated based on their weight, size, magnetic signature (alloy composition), depth and width of edge serrations, and the depth of embossed image.

Conductivity: Being of a different size, being made of different 'paper', and containing different types and numbers of security devices, each banknote of a certain currency and denomination has its own electro-static conductivity and therefore resistance[68]. It is possible, therefore, to tell one from another by running an electrical charge through it which miniature transducers on either side of the note then measure. This method is entirely accurate and works even when notes are wet.

Cash isn't free.

Cash is one of those things – like water out of the tap – that we perceive to be free, but it isn't; there are huge costs involved, from printing it, to getting it into our wallets and purses, and then returning it once we have spent it. Beefing up security, ensuring data protection, and upgrading to new technology does not come cheap. Yet everyone wants – even expects – simple and reliable ways to access their money, even when travelling abroad. When travelling abroad, we often don't know how much each transaction has cost until we get home to read our bank statement. We are usually horrified when we do.

---

[68] Resistance tells us how hard it is for an electrical current to flow. In the quantitative sense, resistance can be defined as the voltage difference between two defined points. Electrical resistance is the inverse of electrical conduction.

One winter not long ago, when renting an apartment in the French Alps, my landlady rang to say she had forgotten to pay her electricity bill and I was about to be cut off. Motivated by the sub-zero temperatures and an acute instinct for self-preservation, I offered to pop down to the utility company's office and pay the outstanding amount of €350 right away. The very charming mademoiselle at the front desk said she would stop the imminent disconnection there and then, but only if I paid in cash. As there was an ATM across the street, that didn't present too much of a problem. For some reason, my debit card wouldn't work, so I was forced to use my credit card. The ATM was not part of my bank's network, so I knew this would increase the transaction cost, but I didn't know how much. I had little option; I needed the electricity turned back on.

A few days later I found out the worst: Knowing that an average ATM debit-card transaction in Europe costs the deployer €0.91 to process, I was staggered to find that, at €39.71 – equivalent to over 10% (11.3%) of the sum dispensed – this single transaction had cost me nearly forty times that. This was made up of three charges, all of which were invisible to me at the time:

- Non-sterling transaction fee of 2.75% of the total transaction = €10.13
- Cash handling fee of 3% = €11.06
- Foreign Exchange £ : € (based on 'bid-offer' spread of 1.27 versus that day's LIBOR of 1.34) = €18.52

Every bank account holder and ATM user has had an experience similar to this, which is why newspapers such as *The New York Times* can print editorials which say things like, "In recent decades, banks, credit card companies and other lenders have made enormous profits from excessive fees and charges."[69]

Most of us don't resent paying a small fee to get hold of our money because we realise that the supply chain is complex and therefore expensive. What we object to is paying a fee we consider to be excessive. Sophisticated bank ATMs cost anywhere between $25,000 and $35,000 per year to run, including capital depreciation[70]. Recouping these costs via a $3 surcharge means each ATM has to dispense cash at least 32 times every day if it is to pay for itself.

In 1996, a change occurred in the law that led to the credit card providers lifting their bans on ATM surcharges. Now anyone with a retail location could own and operate an ATM, and be compensated for their time, trouble and expense. Surely, if I find myself short of ready cash and have to use one of them, then I should expect to pay a surcharge for the privilege?

For a while there was political opposition to surcharging, but the principles of market competition soon prevailed on the simple basis that you can either choose to use an ATM, or you can choose not to. That's the theory anyway, and as long as there is no monopoly or 'closed market' cartel, user fees can be levied at whatever rate the proprietor wishes[71]. It is clear today that most of us have decided ATMs are an essential tool in our lives, and that paying a surcharge – even through gritted teeth – is a small price to pay for the convenience of having safe access to our money 24/7.

Given all this surcharging, you might be surprised to learn that, as a rule, large banks actually lose money on these supposed moneymakers at a rate of about $250 a month per machine. They

---

[69] New York Times editorial, 6 May 2016
[70] ANZ Bank: ATMIA conference, London, June 2013
[71] In November 2016, the US Supreme Court upheld a series of complaints against various credit card companies for 'rigging the market' in just this way.

are, ironically, loss leaders, since banks don't generally charge their own customers if they use the banks' machines. At Bank of America, for example, whose collection of some 16,000 machines is the largest among financial institutions in the US, 85% of all ATM transactions are conducted by their own customers ... about half of whom keep their business with the bank, they say, for just that reason. Wells Fargo has come to much the same conclusion. "If you're looking at it from a pure accounting perspective, it looks like you're losing money," says Jonathan Velline, who heads up ATM banking for the San Francisco-based bank. "But the truth is, if I didn't have ATMs, I wouldn't have customers."

That's essentially what a 2011 Visa survey concluded too, when it showed that 92% of respondents considered convenient ATM access a critical factor in choosing a bank. Or take the Harris Interactive survey, which found that a healthy majority of respondents consider having access to an ATM more important than having access to e-mail.

"ATM fees drive me crazy!" [72]

When not using ATMs from 'our' bank's network, we resign ourselves to having to pay a fee. This can range from anywhere between $2 and $10 per transaction, regardless of how much money is being withdrawn. In today's digital world, when technology is reducing transaction times and costs, these fees don't appear to have reduced much ... in fact, they have more than doubled over the past ten years. With frustrated customers referring to this practice as "legalised extortion," what's actually going on?

Imagine you are making a $100 purchase with a credit or debit card. For that $100 item, the retailer receives about $98. The $2

---

[72] Sales poster, TD Bank, Brooklyn, November 2016

balance gets divided up. About $1.75 (or 1.75%) goes as an interchange fee to the card issuing bank, 18 cents goes to whoever's logo is on the card (say, Visa or MasterCard), and the remaining 7 cents goes to the retailer's merchant account provider as a 'passthrough' fee.

Interchange is a term used in the payments industry to describe a fee paid between banks for the acceptance of card based transactions. Usually the retailer's bank (the acquiring bank[73]) pays the fee to a customer's bank (the issuing bank). In the case of ATMs, it's the other way round, with fees paid by the card-issuing bank to the acquiring bank, supposedly to pay for the maintenance and replenishment of someone else's machine.

In recent years, these interchange fees have become a controversial issue. Regulators in several countries have questioned the collective determination of these interchange fees as potential examples of price-fixing, claiming that they are much higher than necessary. This was certainly the view of the European Parliament who in March 2015 agreed to limit interchange fees to 0.3% for credit cards and to 0.2% for debit cards. They also argued that, because these fees are hidden, consumers are unable to weigh the benefits and costs associated with choosing a particular form of payment.

It gets worse. As my story at the beginning of this chapter makes clear, getting at your cash when abroad can be expensive,

---

[73] An acquiring bank is a bank or financial institution that processes credit or debit card payments on behalf of a merchant, with whom it has a contract. The term acquirer indicates that the merchant accepts or acquires credit card payments from the card-issuing banks within an association such as Visa or MasterCard. The arrangement provides the merchant with a line of credit. Under the agreement, the acquiring bank exchanges funds with issuing banks on behalf of the merchant and pays the merchant for its daily payment-card activity's net balance ... that is, gross sales minus reversals, interchange fees, and acquirer fees. The acquiring bank charges a fee to cover handling and the risk that the merchant will remain solvent.

even when using a debit card[74]. If you need to make an ATM transaction when abroad, in most cases not only will you have to pay the interchange fee but an additional 'foreign' handling charge of up to 2% of the value of the sum withdrawn. You will probably be paying commission of as much as 7% on the 'bid-offer' foreign exchange rate, too. Foreign exchange transactions conducted at the ATM are carried out using a protocol called Dynamic Currency Conversion (DCC). Profit margins on DCC transactions provide a significant income stream for the banks, especially from those machines at locations where there is a reliable flow of overseas visitors such as airports, city centres, tourist attractions, and key bus and rail interchanges. Ramping up the charges at such prime sites sees 10% of the ATM network delivering over 70% of total revenue.

10% of ATMs deliver 70% of total revenue.

And then there is the cost of the systems which ensure all these independent bits of technology function seamlessly together. Collectively, these systems are referred to as 'cash management'. Managing an ATM network entails:

- ensuring availability of sufficient cash
- validating its nominal value
- timely retrieval of funds deposited
- transportation
- sorting
- verifying the authenticity and fitness of banknotes

Despite the rising popularity of internet banking and mobile banking services, the number of ATMs continues to grow

---

[74] When using a credit card, it is probable that interest, typically charged at an annual percentage rate of 27%, will be levied on top of these costs.

worldwide, spurred on by consumers' insatiable appetite for 24/7 cash services. For some years now the terminals have been playing an increasingly key role in banks' interactions with their customers, providing a wide range of services beyond simple cash withdrawal. The machines are no longer 'dumb' like the earlier models but getting 'smarter' and more sophisticated every day, not just in the range of services they provide, but in their ability to forecast, reconcile, and even spot aberrant behaviour. Some even provide a direct communications interface with a human 'bank teller' 24 hours a day, 7 days a week.

Today the global ATM footprint of over three million machines yields more than six billion transactions every year. All this dispensed cash has to be sorted, counted, delivered, protected from thieves and online fraudsters, shielded from the elements, monitored, and reconciled — before it ever lands in your hands. An intricate web of manufacturers, transaction processors, maintenance companies, installers, software providers, and dozens of other ATM-related business segments has risen to meet the demands of this burgeoning industry. It is conservatively estimated that this infrastructure is a $100–150 billion market worldwide, and an industry employing over 1.5 million people.

In 2008, cash accounted for 78% of 388 billion retail payments across Europe, equating to nearly 301 billion transactions that year. The total cost of accepting, distributing, managing, handling, processing, and recycling cash at the time was €84 billion, equivalent to 0.6% of Europe's gross domestic product, or €130 per person per year.

Each ATM in Europe costs roughly €20,500 per year to run[75]. When the opportunity cost of interest lost by cash sitting idle inside an ATM is included, one third (33%) of these running costs are related to the cost of cash itself; a quarter (25%) relate to the

---

[75] Diebold/Securitas, 2013

processing of information, including biometric data; and about one fifth (18%) is spent on maintaining the machine. Based on an 8-year operational lifespan, these figures also suggest an average ATM costs €22,500 to manufacture and install.

As quickly becomes obvious, substantial resources are required to maintain the continuous operation of a secure ATM infrastructure at such a scale. The process involves payment service providers (PSPs), cash-in-transit (CiT) companies, and national central banks (NCB's) providing various services in the cash cycle, which rely on frequent, labour-intensive interventions.

Efficient incident/malfunction management is needed to ensure maximum uptime, a priority for every financial institution. A remote performance monitoring system scans how an ATM's components are functioning in real time and rings an alarm bell whenever any single component stops working. This automated alarm alerts a virtual technician which assesses criticality, and, on the basis of this assessment, decides whether or not to dispatch a specialist human repair team to rectify the fault. This assessment will consider not just the criticality of the failed component – one failed sensor might not be enough in itself to stop the ATM as a whole from carrying out its function – but where and when it took place. Calling out a repair team in the middle of the night is an expensive thing to do, and it might be more cost-effective in some cases to delay their arrival. When implementing a comprehensive device management solution, the need for an on-site visit by a repair team reduces by 20%.

All ATMs have unique characteristics depending on where they are located, and the lifestyle of those who use them. An ATM at Billingsgate fish market in the City of London, for example, reaches peak demand when suppliers have to be paid *i.e* at around five o'clock in the morning. In general, as the graph overleaf shows, if the machine fails during a period of peak demand, the impact will be up to twenty times greater than if it happens during a period of low demand. On average, this occurs in the early hours of the morning when only a handful of people will be inconvenienced.

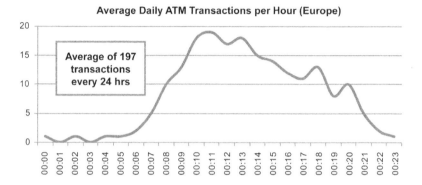

Average Daily ATM Transactions per Hour (Europe)

Automated maintenance schedules also remind technical teams when it is time to replace critical electro-mechanical components before they wear out. This is done with helicopters too, as it's a little late to find out that the 'Jesus' bolt attaching the rotors to the drive shaft have failed once you're airborne. This drive to avoid failures by not letting them happen in the first place supplants reactive support services with intelligent prevention and ensures greater machine up-time. Apart from the potentially long-term negative impact on its reputation, a bank will consider any down-time as a lost opportunity to generate short-term revenue. Banks that can quickly resolve problems not only avoid customer disappointment, but also improve their bottom lines.

Another way to reduce operating costs is to use integrated artificial intelligence algorithms to carry out the complex tasks of cash-demand forecasting and supply optimisation.These algorithms are capable of processing large amounts of information in real time, evaluating patterns, relationships, and trends, making projections, and then choosing the optimum actions from millions of possible combinations ... something the human brain and standard analytical tools simply cannot accomplish.

ATMs located near football grounds on match days, for example, will experience a surge in demand, as will those in airports on certain days of the holiday season. Based on historical and real-time usage data, cash replenishment forecasting models will predict the precise moment a refill is required. An empty machine is just as bad for business as is one full of idle money that nobody wants.

The precision of these complex algorithms can be demonstrated and quantified. A recent study by a company[76] specialising in decision management and artificial intelligence technology applications found that a network of six thousand ATMs can save 22% in logistics costs, 17% in money-handling costs and up to 55% of interest-related costs through applications of such 'intelligent' systems. Expressed in monetary terms, this saves over €1,800 per machine per year, equivalent to €11 million ($14.5 million) per year overall for the operator. The key is choosing technology that supports high levels of both responsiveness and efficiency. Such a solution will incorporate demand-management tools, optimisation engines, complex event processing capability, role-based dashboards, decision management technologies, performance measurement analytics and, most importantly, real-time (or nearly real-time) data.

Cash handling is getting smarter too. As has already been demonstrated, the increased use of so-called 'intelligent' cash cassettes is reducing the need for greater physical protection and cuts transport costs, as cash is protected all the way from the printer, bank, or cash centre to the end-user without the need for

---

[76] Fobiss BV

armoured trucks and multiple security guards. The system is also smart enough to allow the ATM to 'know' where it is, who is handling it, how much cash it is holding, and in what currency and denomination, thereby simplifying and speeding up the process of reconciliation ... which is prone to error when done by humans.

Today, about eight out of ten transactions worldwide are handled in cash. But cash handling and the logistics involved in moving it around is expensive and costs more than $300 billion every year. The lion's share of that expense is incurred in retailing and at retail banks where counting and sorting cash is manual and therefore time-consuming and expensive. According to the European Payments Council, the European Commission's body responsible for cash management matters, it costs every single person in Europe €130 (£94 / $140) per year to have banknotes where we want them, when we want them. As the number of banknotes in circulation is growing, and as more and more banknotes are stored and processed in cassettes, this cost is increasing year on year.

Although consumers spend very little time wondering about how cash miraculously pops out of the ATM, banks spend a great deal of time worrying about it, and are constantly monitoring and trying to predict the amount of cash needed. Why? Because not to do so makes it difficult to optimise costs.

The consequences of mismanaging an ATM's cash flow go beyond running out of money. For a start, such outages fuel the customers 'resentment index' and reduces loyalty. More importantly for the bank, cash management woes such as the failure to correctly predict cash-flow results in the branch over-ordering and thereby holding excess currency. This, in turn, drives up transport costs and increases risk of it being stolen. It also means reduced income as cash lying around idle is not earning interest. Often, there is too much cash at one branch, and not enough at another; a situation that is steadily being rationalised by computer algorithms ise known as 'Integrated Currency Management' (ICM) systems.

Cash forecasting technology creates a central oversight of cash-flow, allowing the amount of cash needed in each branch to be accurately predicted. This really matters when special events or holidays have been incorrectly factored into the cash management equation. Getting this cash-flow right by minimising the number of resupply trips made and optimising the routes taken can reduce the number of branch deliveries by over 40% – from 7,500 to fewer than 4,500 per month -- for a typical national retail bank.

As has been demonstrated, substantial resources are required to maintain the continuous operation of a secure ATM infrastructure. But things are changing. First, innovation has provided more choice over how we pay for things. Second, technology has changed how cash moves through the cycle, including how we access cash and how it recirculates. And third, the notes themselves have become more sophisticated. Technology, in other words, has enabled an entirely new approach to the way cash is managed for society.

As cash handling is both a system and a cycle, there are a number of components which need to be considered as part of an integrated whole. As far as the ATM is concerned, there are five:

- the banknotes themselves
- the way in which they are stored and carried around safely
- the way in which their quality is controlled
- the way they are dispensed
- the way they are recycled.

These components are handled respectively by a combination of central bank, retail bank, private cash centres, and armoured trucking companies ... who are known in the business as 'Cash in Transit' (CiT) companies, and who do much more than move money around. The ATM is now at the centre of this complex supply chain as well as being part of the anti-fraud architecture of modern financial services.

The Cash Cycle

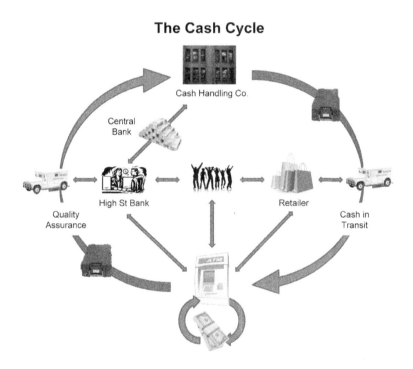

As with our airplane baggage – where suitcases sometimes mysteriously end up at the other end of the planet despite being clearly marked – any technology which reduces the chance of human interference is going to decrease losses. As we have seen, so-called 'smart' cassettes have now got so clever that banknotes can be transferred from one to another without the cassette even being opened. Because the cassettes are tamper-proof, contain

GPS locators, and have microchips embedded in them which records exactly what has happened to the banknotes it either contains or used to contain, a fairly fragile plastic container suddenly becomes extremely secure. Thanks to this traceability, cash centers and commercial banks now have completely new and efficient methods for cash recycling.

Given that the cash itself can be read by an ATM's sensors to determine which currency it is, of which denomination, its condition, and whether it might be counterfeit or not, this even removes the need to manually check for suspect notes. The ATM just took on the role of the central clearing bank, and does so more efficiently than before. Just as smart baggage tagging has done with your airplane baggage.

Instead of having to go through cash centers to be sorted and reconciled, banknotes fit for circulation can now remain directly in the cash cycle. And that's not all:

ATMs that accept deposits allow significant cost savings to be made through increased efficiency and reduced risk. And it's not just everyone involved who stands to benefit, the whole economy has lower social costs to bear through:

- improved liquidity
- reduced labour costs
- reduced banking fees
- enhanced counterfeit detection
- same-day reconciliation
- more accurate records for dispute resolution
- reduction of the robbery threat
- decreased back-office processing time
- complete visibility of where money is in real time

Despite these advances, ATMs nevertheless have to be regularly services and replenished with cash. This is generally undertaken in two ways; either by a cash-in-transit company, or by a merchant in a process known as "self-fill". The CiT company will remove the empty or almost empty cassettes and replace them with full ones. The returned cassettes are then refilled centrally at a cash processing centre. When self-filling, the merchant will refill the cassette with cash from the till as often as required.

In order to meet increasing customer expectations, build brand loyalty, and get maximum return on investment, ATM uptime is critical. For this, information on the status of each and every component in an ATM is needed in real time. The aviation industry works on the same principle: each engine on a commercial airliner tracks its own performance and sends a signal to the manufacturer every fifteen minutes or so which enables remote maintenance teams to be alerted to upcoming faults before they occur. Using such systems, maintenance teams can be dispatched from a central location to meet a specific plane before it lands rather than hang around multiple airports on the off-chance that something might go wrong, which is a much less efficient and more costly approach.

As the range of equipment available for accepting, re-cycling and processing notes increases, so does the need to capture the information from these new systems. The information required for the management of note stocks, their location, and the quality of the currency in circulation is critical to the effective management of cash for central banks, commercial banks and cash management companies alike. Furthermore, this information needs to be presented in a consistent manner which allows managers to assess note quality, equipment performance and evaluate trends, so that issues can be addressed before they become critical.

Fewer faults and faster fixes increases efficiency. Faulty ATMs don't just cost money but erode customer confidence. Remote monitoring technology allows an administrator to oversee the ATM network and identify problems ahead of time. Predictive maintenance is part of this process and allows insight into issues before they occur, thereby saving time and money as it minimises – and often eliminates – the need for site visit by a technician. Remote fault monitoring triggers dispatch of an engineer with the replacement part and correct tools, many of which are specialised and can only be operated by certain individuals. No single individual can maintain all components, which means that no single individual knows exactly how each component interacts with the other. Just as with helicopters, intelligent maintenance tracks the lifespan and/or use of critical components and replaces them before they fail.

While all this is going on, cash handling systems are forecasting how much cash should be lying idle in the machines – and therefore out of circulation – at any one time, and when. This is based on interest rates and current and historical transaction records. Such insight allows for just-in-time cash replenishment without compromising cash availability. Fully integrated and interoperable systems like these also reduce operating costs through economies of scale.

While technology is fostering ever-more efficient cash management, the cash itself remains more or less as it was. Even polymer notes wear out. So, what happens when they do? Where does money go to die? Is it illegal to deliberately destroy it?

According to the Bank of England, it is a treasonable offence under the Currency and Bank Notes Act of 1928 to deface the image of the Monarch by deliberate mutilation of a banknote. The 'Fed' is only slightly less severe when it suggests in Section 333 of the 'national mutilation of bank note obligations' ordnance that

"Whoever mutilates, cuts, defaces, disfigures, perforates, or unites or cements together, or does any other thing to any bank note, draft, bill, or other evidence of debt issued by any national banking association, or federal reserve bank, or the federal reserve system, with intent to render such bank bill, draft, note, or other evidence of debt unfit to be reissued, shall be fined under this title or imprisoned, or both."

Luckily, despite the drastic language, the US Mint confirms that the law only prohibits you from destroying or defacing money if you still intend to use it as money. If you're destroying it for some other reason – to dye it or spray it with superglue to stop criminals from running off with an ATM cassette, for example – it's legal. That's how the companies that make those souvenir coin-flattening machines are allowed to continue operating. But it's a serious issue. Have you ever tried paying for something with a particularly worn and tatty banknote and found yourself wondering just how many hands it's passed through? Bankers do and go to great lengths to track how fast money moves around the economy – they refer to this as "the velocity of money" – because it's their job to make sure that what you're trying to pass off as legal tender can be recognised for what it is and doesn't wear out.

But recent trends show that we are handing over fewer banknotes than ever during a single transaction, with the average value of each transaction down to £9.47 in the UK. With fewer cash transactions taking place, banknotes should last longer, with polymer notes lasting up to three times longer still.

A banknote's longevity depends on a variety of factors, among which are climatic conditions, handling methods, frequency of use, and the replacement rate. In developed countries, they are scanned regularly to check that they're not worn out or counterfeit. With an average lifespan of slightly less than two years, £5 and €5 notes change hands about 258 times, while £10 and €10 notes, being used less often and therefore having a lifespan of nearly three years, change hands 594 times before being withdrawn from circulation and disposed of.

This is where the expression 'cash recycling' takes on another meaning. Expired or worn-out banknotes are taken out of circulation by the banks or cash centres, cut into small pieces, and eventually end up as fuel for power plants or as agricultural compost ... meaning that the phrase "money doesn't grow on trees", when reversed, is at least only partially true. Cash can be recycled in other ways, too. A few years ago, Dublin-based artist Frank Buckley built himself a house made of $1.4 billion in shredded Euro notes with each brick made up of 40,000-50,000 banknotes.

A digital society does not necessarily mean a cashless society.

While it might be clear what physical currency looks like, there seems to be some confusion between what constitutes e-money, mobile money and a digital wallet.

ATMs convert electronic digital currency – what is commonly referred to as 'e-money' – into its physical form. In this, it's just part of the payments and cycle. Nowadays, however, cash is not necessarily needed in the cycle and e-money can either be transferred from a stored-value card where embedded microprocessors have been pre-loaded with a finite monetary value or transferred electronically, usually over the Internet. Bank deposits, electronic funds transfers, and digital currencies are all forms of e-money.

E-money is either centralised, where there is a central point of control over the money supply – or decentralised, where the control over the money supply can come from various sources.

Decentralised e-money is a form of digital currency. Confusingly, there is a difference between e-money and a digital currency in that e-money doesn't change the value of the fiat currency it represents. In other words, all digital currency is e-money, but e-money is not necessarily digital currency. More confusingly still, crypto-currencies such as Bitcoin allow e-money systems to be decentralised.

Simply put, the Bitcoin system serves as a distributed digital ledger. Records of transactions – the 'blocks' – are added to a central list – the 'chain' – of all transactions. This is referred to as 'the blockchain'. Complex algorithms and encryption are applied, with copies of the blockchain kept in many places. This enables multiple players to arrive at consensus over which transactions happened, for how much, and when. Forged entries will not match up with the other versions and therefore will not be accepted by the payments system. "It proposes to be a platform of truth and trust which could replace the systems that exist today," says Rahul Singh, president of financial services at HCL Technologies. As Philip Stafford of the Financial Times put it in an article on 26 September 2016 when attempting to explain the challenges of distributed ledger accounting which lies at the heart of the blockchain concept, "The block chain aims to combine the peer-to-peer computing ethos of silicon Valley with the money management of Wall Street, automating the network of trust on which modern finance sits."

Although blockchain technology has great potential, financial institutions are still working out exactly how it can be used to thwart cyber-crime, however. Two main issues remain to be tackled before payments in crypto-currencies become widespread. The first is that the blockchain is decentralised which gives rise to a whole host of issues around security and governance. The second challenge lies in standardising the protocol to be used.

Interestingly, central banks are belatedly attempting to harness the technology ... which is ironic given that Bitcoin was created by libertarian programmers with a deep suspicion of central banks and of the national currencies they issue. For the central banks, the promise of the technology is that it would allow them to track every pound, dollar or renminbi on every step of its journey through the financial payments system in real time, something that is apparently impossible now. The official goal is to make the wholesale financial system of clearing and settlement more transparent, fast, efficient, and secure. But there is another reason the blockchain concept is being pursued: much less money would be left sitting idle while banks reconcile their different ledgers, as happens now. This win-win potentially means more revenue for those involved while at the same time improving gross national income statistics.

Switching to a blockchain-based model also has the potential to cut billions of dollars of hidden costs out of the financial system by making global payments, trade finance, syndicated loans, and equity clearing more efficient. The trouble is, the technology is overshadowed by questions about how it can be made both secure and fast enough for large financial institutions which still need to comply with rules about customer authentication, money-laundering, and counter-terrorism. As explained by Michael Thomas, a partner at law firm Hogan Lovells, clearing is complex. "If title to an asset is to be transferred using a blockchain system," he says, "it's also important to have a system for the transfer of any corresponding cash payment for that asset ... otherwise any benefits in reducing counterparty risk from the blockchain will be imperfectly realised."

In addition, electronic currencies can be hard or soft. Hard e-money does not have services to dispute or charges to reverse. In other words, it is akin to cash in that it only supports non-reversible transactions. Reversing transactions, even in the case of legitimate error, unauthorised use, or failure of a vendor to supply goods is difficult, if not impossible. The advantage of this arrangement is that the operating costs of the electronic currency system are greatly reduced by not having to resolve payment disputes. Furthermore, e-money transactions clear instantly, making the funds available immediately to the recipient. Bitcoin is an example of such a 'hard' system.

Soft e-money on the other hand, is a system that allows for reversal of payments, for example in cases of fraud or dispute. Such reversals have a clearing time of 72 hours or more. PayPal is an example of such a 'soft' system.

In general, a digital wallet is neither a piece of leather containing credit cards and cash nor a smartphone. Rather, it is a software application on a mobile device that serves as an electronic version of a physical wallet. ApplePay is an example of a digital wallet which uses near-field communications and security methods like tokenisation to create a user-friendly, mobile and ostensibly secure payment experience. Being Internet-based and slaved to inbuilt GPS software, tomorrow's digital wallets will do more than allow retail payments but will enable consumers to receive targeted promotions as they approach a store.

Near field communications (NFC) technology allows you to wave your smartcard or smart phone and make a payment up to a certain limit without any further verification. This convenience, once it catches on, will be hard to beat. Trouble is that retailers have yet to change their existing point-of-sale terminals and the systems that go with them in order to accommodate this new payment technology at the scale required to gain universal acceptance. If you think ATM manufacturers are slow with product development and adoption, retailers are worse ... mostly because it is them who bear the cost of installation, the costs of transacting, and the risk when things go wrong.

Those who have a lot to gain from widespread public adoption of digital payments are quick to point out that the process of processing a digital transaction through modern encryption technology is safer and far more efficient than using an ATM to withdraw cash or using a card, especially a card which has not been EMV (chip and pin) enabled.

The biggest risk isn't that of data interception by a nefarious third-party, but physical loss of the device containing your mobile wallet application. In the time it takes for you to realise you have lost your smart phone and found some way of reporting it stolen, a thief could have used it many times. Of course, this is all easily preventable with a strong password or the use of a biometric scanning device. All of which is fine in principle. In practice, however, if we can remember the robust alpha-numeric passwords we are supposed to change regularly, we find entering passwords a mental pain in the ass. Instead, we prefer the fingerprint technology now in common use in smartphones. But even these are not as secure as we would like them to be, all of which is a risk we might be willing to bear when buying a cappuccino but not when transacting for what we consider to be significant amounts of money.

Another risk involved in digital wallet adoption is one that has yet to be satisfactorily answered: who accepts the burden of personal liability in the event of fraud? Most credit card companies currently shoulder this risk, but this will soon not be the case when the card is tied to a mobile wallet.

Right now, there is a great deal of fragmentation in the payments industry, and until credit card companies, digital wallet technologies, and consumers get their goals aligned, it is likely to be a long and winding path to gaining the levels of consumer trust and retailer adoption needed to complete the digital payment ecosystem.

What is also unknown is how human behaviour will react to what some, especially older generations, perceive to be invasions of their privacy. Also unknown is the extent to which cyber-crime will defraud mobile users, and the user's reaction once they realise their phone has disappeared and with it, their virtual cash.

chapter six

political economy

: ~' /pə'lidək(ə)l.-'/ē'känəmē

adj.,noun
political policies and economic processes; their
interrelations and influence on social institutions

FROM ITS HUMBLE AND UNCERTAIN BEGINNINGS fifty years ago, the ATM has become so everywhere that it's almost impossible to envisage life without one nearby. The near-total global integration of ATM networks means that we can travel almost anywhere in the world with just a piece of plastic in our pocket, confident that we'll have access to cash in places as far afield as the Australian outback, the tundra of Canada, Red Square in Moscow, Easter Island, and even Antarctica. This was unthinkable twenty-five years ago.

Because its very ubiquity means it has to work in some pretty extreme and remote settings, the ATM has to be versatile as well as robust. As a result, not only can they be found in the most unlikely of places but are also sometimes used in the most unusual of ways. There is an ATM at Whistler Mountain ski resort

189

in Canada with a heated facia that allows skiers to warm their hands while collecting their money ... and all without having to take their skis off. There is a nother in Los Angeles that dispenses made-to-order cupcakes[77]. There are even ATMs in Dubai that dispense gold bullion.

One particularly robust ATM remained standing after Hurricane Katrina swept through New Orleans in 2005 while the entire two-floor marina chandlery in which it was installed was washed away. Not only did it remain standing, but it was fully functional within 24 hrs of the flood waters receding.

Some ATMs have to be able to cope with the cold, others with the hot. Antarctica's ATMs are located inside McMurdo Station, the USA's Antarctic science facility near the South Pole. There are also ATMs north of the Arctic circle on Baffin Island where temperatures of down to −26 °C (−14.2°F) and 'white out' conditions from blowing snow are very common. In northern Finland's Lapland, ATMs have been upgraded to keep the cold and snow from storming in by blowing hot air out of the dispensing slot.

The Tjuntjuntjara Aboriginal Community in Australia's outback falls within the boundaries of the Great Victoria Desert Nature Reserve, 550kms east of Kalgoorlie-Boulder, and over 1,000 kms from Perth, way beyond the reach of any mobile services. Here in the remote desert, accessible only by Aboriginal tracks, and where camels and other exotic animals can sometimes be spotted, it is not unusual for temperatures to reach 50°C (122°F). A mail plane comes once a week on a Wednesday and a truck only occasionally for essential supplies. The people who live there are known today as the Spinifex people, traditional custodians of 55,000 square kms of Spinifex native title

---

[77] These are, in fact, vending machines although they are referred to as 'ATMs' by those who own and use them.

determination area, for which permission is required to enter. The ATM is located in the general store of the town. It does in excess of 500 withdrawals and 200 balance enquiries per month, every month. It uses power generated by solar panels and transacts via satellite uplink.

At 7,800 ft above sea level, the ATM located at Sust along the ancient Silk Road in one of the remotest parts of Pakistan's Karakoram mountains is reputed to be the world's highest ... though, typically, this is contested by both India and China. The world's lowest ATM can be found in Israel at the Dead Sea resort of Ein Bokek. The ATM, installed by a local grocer, sits near the beach at 1,380 feet (421 m) below sea level. In Israel, ATMs are commonly referred to as 'Kaspomat', a Hebrew term meaning 'automated money'. As there is reputed to be an ATM on at least one of the US Navy's nuclear-powered submarines, this record might get broken from time to time, however, by the world's deepest.

The small town of Fuerte Olimpo, with about 5000 inhabitants, and located 800 km from the capital of Paraguay in poorly accessible conditions, now boasts an ATM which is serviced by technicians who have to fly in on one of two weekly flights available. Before the ATM was installed, the 5,000 villagers had to travel 300 km to use the nearest ATM.

Even places of great historical antiquity and mystery, such as the Forbidden City in Beijing, China, are not beyond the reach of today's ATMs.

ATMs are becoming more and more common inside churches, too, where 'offerings' are increasingly made at the ATM before, during or after the service. There is even one in Vatican City in Rome. Since Vatican employees and diplomats accredited to the Holy See must be able to speak and write in Latin, the Vatican houses the only ATM in the world to offer Latin as a language option. The welcome screen reads, "Submitto scidulum quaeso ut faciundam cognoscas rationem" which in English translates to, "please insert card to access permitted operations". When finished, users are bid farewell with a cheery, "Carus exspectatusque venisti" which more or less means, "You're welcome. Come again soon." Perhaps unsurprisingly, the Vatican's ATM is also one of the world's most linguistic as it offers instructions in almost every European language as well as Russian and Arabic.

More idiosyncratic still are the ATMs to be found in Buckingham Palace, inside private jets ... and in the back of the rock star Alice Cooper's car[78].

ATMs play a critical role in disaster response operations.

To the owner and user, these ATMs are far from being outliers. Some are even taking on an increasingly pivotal role in humanitarian disaster response operations. During flood relief operations in Pakistan, ATMs were flown in underslung from helicopters to help distribute much-needed cash to survivors. So remote have been some of the places in which ATMs have been needed to complement disaster response operations that they have been attached to donkeys in Baluchistan, elephants in India, boats in Bangladesh, and pick-up trucks in Sudan's Darfur. They are proving invaluable in other disaster settings, too. Iris-recognition ATMs are used throughout the refugee camps of Kenya, Lebanon, Jordan, and Somalia. In Turkey, pre-paid debit cards are used at ATMs to pay stipends to exiled Syrian teachers.

Cash works. It enables disaster survivors to spend money on things that accelerates their recovery and improves their lives far

---

[78] For more on idiosyncratic 'extreme' ATMs, see the list on page 315 or visit www.lovecashmachines.org

more cost-effectively than by providing in-kind donations. And, contrary to popular belief, recent research by The World Bank and Harvard University shows it doesn't cause them to go out and squander it on alcohol or cigarettes. Cash is increasingly being recognised as a means of providing people with essential disaster relief quickly while at the same time stimulating local economic recovery. When not conditional on, say, school enrolment for the children or attendance at a training session on how to build a disaster-resilient home, it also affords flexibility and dignity by allowing recipients to choose what they want to buy, when they want to buy it. The UN World Food Programme's shift from food aid to food security – which adds vouchers and cash to its collection of 'hunger solutions' – is one indicator of the foothold that cash transfers are beginning to have in humanitarian response. Similar trends can be seen in the shelter sector, particularly for rental support and, in colder climates, subsidies for winter fuel.

An estimated 6% of the $24 billion annual global humanitarian aid budget is currently dedicated to what the aid industry calls cash transfer programming in disaster settings. This is 'aid-speak' for the handing out of cash rather than more traditional distributions of in-kind relief commodities such as food, buckets and blankets. There is an appetite amongst donors such as the UK Government's Department for International Development for such programming to scale, with a tenfold rise sought by 2020. The United Nations' refugee agency, UNHCR, has committed to doubling the volume of their cash based interventions by 2020 on the basis that refugees lose their ability to earn and spend when forced to flee their homes. They justify this approach by arguing that the flexibility that cash and vouchers offer makes for a more dignified form of assistance and gives refugees the ability to prioritise and choose what they need.

"The use of cash-based assistance has been a real game changer in the way we help refugees. We have now decided to make it a worldwide policy and expand it to all our operations, where feasible*," said the acting High Commissioner, Filippo Grandi, in October 2016.

More importantly, perhaps, cash-based interventions make the displaced less likely to resort to harmful coping strategies, such as survival sex, child labour, family separation and forced marriage. They also directly benefit the local economy and can contribute to peaceful coexistence with host communities

Where access allows, ATMs play a central role in enabling these cash distributions. Yet, as the director of cash services at one of the world's largest non-governmental organisations, the Norwegian Refugee Council, pointed in January 2016, "ATMs are central to our digital cash transfer planning but we have limited knowledge of how they work, including in terms of their potential to add social value." The CEO of a for-profit social enterprise, Segovia, expanded on this when he said, "Over $300 billion is given directly to the poor every year, but these programmes often rely on outdated, insecure, and incompatible payment systems."

Cash is the trusted medium of exchange, especially in low-income societies where more sophisticated transfer mechanisms are not well understood; where the technology is unknown; where liquidity can be a challenge; where a regular power supply cannot be relied upon; and where reconciliation systems are based on thumb-prints and pencils. As the 2015 earthquake response in Nepal and response to Hurricane Mathew's rampage across Haiti in 2016 demonstrated, in appropriate circumstances ready access to cash following a disaster can be as important to survival and recovery as physical provision of tents, blankets, food, stoves, cooking sets, and hygiene parcels.

In the days following Typhoon Haiyan's devastating pogrom across the southern Philippines in November 2013, good preparedness planning meant that ATMs were up and running within 24 hours, long before mobile telephone services were restored. A small stockpile of essential spare parts and a few cash cassette's in the safe are important preparedness measures. Equally important for the short-term needs of survivors is the necessity to have mobile ATMs at the ready with their own power supply and satellite communication uplinks.

This adoption of cash transfers as a programme tool to complement the more traditional approach of bulk in-kind distribution is re-engineering the humanitarian world by reshaping business models, forging new public and private partnerships, and changing the relationship between provider and recipient. In the process, it is opening up new revenue streams for the financial services sector.

Overall, cash transfer programming offers substantial benefits for beneficiaries, governments, and financial service providers alike. Government finances will be improved because of the reduced burden on subsidies, reduction in 'leakages', and increased political legitimisation which results when the people see "their money" transferred faster, more reliably, more efficiently, and with less opportunity for corruption.

Slowly but surely – and whether humanitarians like it or not – the aid industry is privatising, and companies that facilitate or provide physical or electronic cash transfers will become increasingly important partners in disaster responses.

An opportunity exists for the cash management industry to provide its expertise, resources and network to build better market understanding and new cash transfer mechanisms that can be readily deployed in different post-disaster settings. But this transformation can only be fully realised if provisions are made well before a natural disaster hits. A report published in May 2015 by the UK's respected Overseas Development Institute noted that " ... in most contexts, cash provides ample opportunity for going to scale ... and there is a greater role for the private sector to deliver cash."

All this having been said, cash transfers are not a 'silver bullet' for eliminating suffering or rectifying the current inefficiencies in aid delivery, but they can go a long way to improving lives and restoring livelihoods in the wake of disaster. Despite this, many in the aid world are unaware of the broad range of societal benefits offered by cash and cling to a number of misperceptions about the usefulness of cash transfers in times of crisis:

Cash Misperception # 1: Cash is redundant. Reports of the death of cash have been greatly exaggerated. Despite political moves to embrace the utopia of a cashless society, cash remains by far the most used payment instrument across the world. This might not be the case for ever, but it is certainly the case now, even in the developed world.

Cash use might be declining in wealthy economies such as the United States, but mobile payments are not growing to fill the cashless gap. This was the surprising finding of a survey conducted by Accenture, a consultancy, which found that cash was still the most used form of payment in the US, but that its use had declined by 7% in the year 2015-2016. Awareness of mobile payments options had increased, but actual use had remained static, with less than one in five consumers adopting the technology.

More than one-third (37%) of the 4,000 consumers surveyed said they were content to continue using cash and plastic for their everyday transactions, with one fifth (19%) preferring not to register payments credentials into their mobile phone over concerns about unauthorised transactions.

Accenture's analysis suggested that any shift in consumers' payment habits will take "more than just another 'me too' mobile payments option", going on to say that "the ability to tap-and-pay is simply not enough in itself". Although mobile wallet use is bound to rise – it is currently used by only 14% of the US population – it is likely that consumers are not making a mass move to mobile money for three main reasons:

- The existing payments system isn't broken
- Concerns about cyber-crime haven't gone away
- There are no incentives to prod consumers into making the shift.

Take the case of the European Union, too. According to the European Central Bank (ECB), over half (59.7%) of all transactions are still made using coins or banknotes. Even in the UK, where non-cash payments overtook transacting using physical currency in 2014, the Bank of England reports that the value of cash in circulation has risen steadily year-on-year by about 11% for the past 20 years i.e outstripping inflation and rises in GDP. According to the US Federal Reserve, the volume of US dollar bills in circulation around the world has also been steadily increasing.

This cash is made available to the general public primarily through ATMs. By the end of 2014 there were over 3 million ATMs around the world, a figure which is predicted to increase to over 4 million by time of the ATM's 50th anniversary in mid-2017. This means a new ATM is being installed somewhere in the world every three minutes.

Cash Misperception # 2: M-Pesa is the mobile money model to emulate. M-Pesa (M stands for mobile and Pesa means 'money' in Swahili) is the iconic mobile banking service that led to copycat businesses around the globe. Essentially, it is an electronic payment and store-of-value system accessible through mobile phones which enables the user to send money in electronic form, store money on a mobile phone in an electronic account, and deposit or withdraw money in the form of hard currency.

In the drought prone area of East Africa, where banks are few and far between, cash programming would once have meant setting up payout centres and the recipients having to travel considerable distances to collect their money. Now, thanks to 'mobile money' services like M-Pesa, transfers can be sent direct to recipients' mobile phones and the cash picked up from their nearest agent. This saves time, increases accountability, and is altogether more secure. It is cheap, because it uses the already existing network of mobile phone agents; it's easy to reach scattered recipients over a wide area – as long as they have mobile phone coverage – and most of the recipients find it convenient and more dignified than coming to a central distribution point to get assistance. Organisations running the programmes no longer have to handle large amounts of cash, either.

Lack of in-depth understanding of how the M-Pesa mobile money model actually works in practice, however, has resulted in a situation which is not necessarily in the best interests of those affected by disaster, nor for those for whom financial inclusion represents their best opportunity of clawing their way out of poverty. This is because most digital mobile money transfer models, including M-Pesa, are not virtual but rely on the use of physical currency. With M-Pesa, it is estimated that over 90% of all mobile payments involve 'cash in' and/or 'cash out' at some point in the cash cycle[79]. Equity Bank in Nairobi estimates that

[79] Dan Littman, Federal Reserve Bank of Cleveland; personal communication, 4 October 2016

98% of the value of all financial transactions in Kenya today are still made in cash. Equally significant is that, though proximity to bank agents in Kenya has increased since 2013, 80% of users still access financial services either through their branch (41%) or via an ATM (38%)[80].

In light of M-Pesa's impressive success and widespread use, it is also perhaps surprising to note that since this mobile money system was introduced, both the velocity (rate of cash use) and volume of currency in circulation in Kenya have actually increased. This partly reflects the practical realities of mobile phone use in low-income countries where electricity provided by solar lanterns can be in short supply; where phones are for the most part not 'smart'; and where ownership is often shared. This is not just the case in low-income societies. Despite Japan's reputation as an advanced mobile payments market, the country still has one of the highest rates of cash transacting in the world. It is also interesting to note that, according to the ECB[81], "Most forms of electronic payments seem to be competing for volume with forms of payment other than cash."

Cash Misperception # 3: Cash is apolitical. Moves for entire countries to go cashless are essentially political acts, and are driven in part by a legitimate desire to increase transparency, reduce corruption, and minimise opportunities for tax evasion. But they are also driven by a less legitimate desire to increase the profits of those involved in digital payments. Countries such as Sweden, Denmark, Rwanda and Nigeria have all declared their intent to go 'cashless' (or at least not force retailers to accept physical currency). To achieve this goal, access to cash first has to be restricted. This is done by introducing legislation that limits the

---

[80] Central Bank of Kenya: Household Financial Access Survey, February 2016
[81] ECB: Consumer Cash Usage – a cross-country comparison; Working Paper Series No.1685, June 2014

use of cash at point-of-sale, reduces the number of ATMs and bank branches, and by allowing charges for ATM withdrawals to remain artificially high.

Bearing in mind Misperception # 1, this tells us is that access to cash is being deliberately restricted in the face of an increasing demand for cash from the general public. It is fair to ask why any industry would actively limit access to a product which is apparently in growing popular demand? This trend will not serve those affected by disaster well, either in the short or the long term. The humanitarian sector should strive to ensure that the full range of payments options are properly considered. Apart from anything else, the 'dignity, humanity and agency' logic of moving from in-kind aid delivery to cash transfers demands that recipients should have the freedom to decide their own priorities and choose their own solutions.

Cash Misperception # 4: Cash is inefficient. There are a number of attributes that distinguish cash from other payment instruments. Firstly, cash is legal tender. That is to say it is a medium of exchange recognised by a country's sovereign and legal system as being valid for meeting financial obligations, both personal and national. Non-cash methods of payment do not comprise legal tender and are therefore 'un-secured'.

Furthermore, non-cash payment methods are not public goods. As Thierry Lebeaux, CEO of ESTA, a trade association, puts it, "Cash is public money generating public revenue, while electronic money is private money generating private revenue." Promotion of non-cash payments leads to the transfer of risk and shifting of costs to the recipient while creating additional revenue streams for all those involved in the transaction.

It is also worth noting that digital money doesn't recirculate, while physical currency does. This means that, while digital money is creating revenue for those involved in the transaction, it exerts no leveraging affect within local communities. According to a study by the UK's Overseas Development Institute published in 2015, every dollar distributed in the form of humanitarian cash transfers, on the other hand, exerts a 'multiplier' effect within the local economy, generating in the region of $2.4's worth of additional transacting.

Cash Misperception # 5: Cash is insecure. Cash is more secure than supposed. The insecurity of cash is relative and dependent on the online insecurities that come with paying electronically. Of course there are disadvantages in carrying cash around, especially where law and order has broken down. Logic would suggest that those affected by disaster might regard the use of cash as less safe than making card payments or paying electronically by mobile smart phone as a result. But they shouldn't, and for one simple reason: the risk of having their money stolen through cyber-fraud is thousands if not hundreds of thousands of times greater than the risk of being mugged, even in a refugee camp.

From the point of view of the merchant – who probably has no safe, and whose sometimes substantial cash holdings are kept in a drawer – holding ready cash involves an even greater risk. Yes, non-cash payments can reduce this risk, but the need for sustained liquidity in times of crisis means that some method of safely transporting and storing large volumes of physical currency needs to be available. Recycling ATMs and 'smart' cassette technology offer these safeguards.

Cash Misperception # 6: Cash is inconvenient. In low-income societies, people are familiar with cash. They trust it. This doesn't change when disaster strikes. For it's just when dazed survivors realise that modern world conveniences have disappeared, when there is no power, no water, and no food, that they turn to cash, and lots of it.

Ready cash is a 'safe haven' that forms the foundation for the way we operate socially as well as economically, especially in uncertain economic times. Cash is simple to stockpile prior to a hazard event occurring and is resistant to systems failures and power outages once it has. Those lucky enough to have early warning of impending disaster – when hurricane or tsunami warnings have been issued, for example – always try to stock up on cash as one of their first priorities if they have the time. ATMs along the Florida and New Jersey shorelines had to be replenished at four times the normal rate as Hurricane Matthew and Super-Storm Sandy approached, and the mayor of Tacloban in the Philippines not only asked the banks to stockpile large quantities of cash but got them to build walls around their ATMs as the storm-surge in front of Typhoon Haiyan approached in December 2014.

The experience of New Zealand's response to the Christchurch earthquake of 2011 explains why cashless payments systems are not applicable in the context of natural disasters. According to Alan Boaden[82], Head of Currency at the Reserve Bank of New Zealand, "Access to physical currency is an immediate priority in times of national emergency, even in a country where 75% of transactions are normally made with electronic payments. In fact, when electronic retail payment systems are not working, electronic payment becomes a vulnerability, not a strength." In terms of lessons learning, he went on to suggest that "local authorities need to work closely with banks and cash-in-transit companies in high-risk areas prior to these types of natural hazards occurring." According to Ecuador's Minister of Finance, it was much the same during Ecuador's response to its earthquake in April 2016. "People are

[82] Boeden: Demand For Cash After a Natural Disaster; Singapore International Currency Conference, October 2011

comfortable with cash," he said, "They trust it, and it was our job to invigorate recovery by making sure they had access to it."

Cash Misperception # 7: Cash is expensive. Banknotes have to be printed, stored, and distributed. All this requires insurance, security, machinery, staff, real-time tracking, and linked up IT systems. However, the ECB concluded in 2015 that "cash is less expensive than electronic payments, both for society as a whole and for retailers." This goes against conventional thinking, particularly by those in favour of mobile payments. "On average," the report says, "cash payments show the lowest social costs per transaction ... while merchants would be better off if transactions currently executed with debit cards were instead carried out using cash."

Cash Misperception # 8: Cash is simple. Conducting cash transfers at scale in a disaster zone is a highly complicated and complex undertaking. There is no one-size-fits-all approach for the optimisation of cash management. The availability of, and access to non-cash payment infrastructure – as measured by the number of debit cards issued, point-of-sales terminals available, or percentage of population with a bank account, for example – differs widely per country. The optimal cash cycle also depends on many country specific factors, including culture, geography, the regulatory environment, and the efficiency with which physical currency is recirculated in society.

Cash Misperception # 9: Cash has few indirect societal benefits. For generations, physical money has served a societal role in educating children on both the value of money and how to manage or save it. For largely psychological reasons, this cannot be replicated by non-tangible forms of payment. Cash allows consumers to exert control over their spending habits. A psychological phenomenon known as the 'dissociation effect' ensures that digital-only consumers over-spend by anywhere between 20% and 40%. This not only has short-term implications for the individual beneficiary but has longer-term implications for aid agency and donor budgets. In addition, humanitarian cash transfers in times of disaster represent not only a more efficient way for donors to donate but also a more efficient way to provide relief to survivors i.e more can be done with less.

When discussing the cost-benefit of any one particular payment instrument over another, the discussion should not be limited to the quantifiable only. Qualifiable elements such as ease-of-use, convenience, access, and availability all need to be factored in, as do other types of non-quantifiable components such as impacts on culture and gender equality.

Cash is also more than a convenient method of payment. For those who are illiterate or for people who find it difficult to manage an electronic budget – and that includes the vast majority of those affected by disasters around the world – cash has a number of important symbolic values: It gives them dignity; it affords choice; and it puts them in charge.

The social impact of the visual imagery printed on banknotes should not be ignored, either. As the later story from Libya makes clear, banknotes are part of the glue of society. As with a flag or a national anthem, they provide a vivid symbol of a country's sovereignty and provide people with a living link to their cultural and historical roots. Many people in society, old and young alike, feel comforted by the presence of physical currency.

Cash Misperception # 10: ATMs are unsuitable for humanitarian action. As has been pointed out already, ATM penetration is increasing in low-income societies, not reducing. Rural ATM networks in the Southern Americas, across Africa and Asia, and from India to the Philippines are expanding. In part this is because governments are keen to stimulate financial inclusion and reduce disaster risk, and have understood the role that ATMs can play in societies in crisis. Like the commercial banks, they recognise that technology – particularly biometric technology – has enabled the ATM's transformation from 'automated teller' to stand-alone 'branchless bank'; that remote operations are increasingly feasible; and that an adequate return on investment is now possible, not just in revenue terms but for society at large.

In another example of the humanitarian role played by ATMs, two hundred thousand residents of the Mathare slum area of Kenya's capital, Nairobi, are now able to access safe water through a network of what they call "Water ATMs" located in central and well-lit areas along the main streets, making them easily accessible day and night. The machines – not really ATM's in the strict definition of the term – have revolutionised clean water availability and distribution to populations that have long been at the mercy of unscrupulous water cartels who charge up to 100 times more for water which is frequently contaminated. The daily income in a slum household in Kenya is just over one dollar, and the average home uses around 100 litres of water a week. The water vendors operating in the slum would charge Sh50 (50 cents) for a 20-litre container of water. With the introduction of the Water ATMs, weekly expenditure on water in Mathare has been reduced from Sh250 ($2.5) to Sh2.50 (2.5 cents)[83].

A public-private partnership recently concluded between Nairobi's Water and Sewerage Company, the city's main water distribution agency, and Grundfos, a Danish water engineering firm, resulted in the installation of fifteen of these Water ATMs. The Nairobi Water and Sewerage Company had been trying to get viable solutions to water supply problems within informal

---

[83] www.bbc.com/news/world-africa-33223922; 22 June 2015 [accessed 3 May 2016]

settlements for years. "Initially, our pipes were vandalised by these same cartels that sold water to residents at exorbitant prices," says Mbaruku Vyakweli, the company's communications officer. "Now, all we need is a safe and secure area, agreed on by the residents, and we supply the water from our own dams and reservoirs. Our prices are constant because the product is available throughout. Plus, the water ATMs are run and monitored by residents; they own them and therefore take better care of them."

To buy clean water, users load points on to smart cards given out to residents for free and loaded with an amount of their choice from a Nairobi Water and Sewerage Company outlet. By a simple swipe of their card on the ATM's sensor, water is released from the main storage tank into a waiting container. The water reaches the ATM through 18km of newly laid pipes connected to the city's main supply.

Residents say the health benefits of the scheme are already being felt. In the summer of 2014, a cholera outbreak swept through the slum and other surrounding areas, resulting in two deaths. The cramped slum conditions coupled with poor sanitation – including unhygienic water – contributed to the spread of the disease. At the time, Médecins Sans Frontières, an NGO, was recording 200 new cholera cases every week. Life has also become much easier on account of the Water ATMs. "I am saving on other costs," says one smiling woman, adding, "Because the water is already treated, I spend less on charcoal or kerosene to boil my drinking water, and can use the money saved to pay for my daughter's school books."

The ATM plays a major role in combatting global poverty.

The ATM's role is not limited to disaster settings and humanitarian intervention; it plays an important role in defeating global poverty, too. An accumulated body of evidence[84] supports the convictions of policy makers who believe that financial inclusion – providing universal access to basic financial products and transaction services – is an important ingredient for social and economic progress, and a vital tool in the global fight to reduce poverty. What this means in general is that a bank account helps poor families improve their lives by providing a gateway to formal financial services.

At the moment, half the world's total adult population – over 2.5 billion adults globally – have no access to formal financial services i.e those delivered by regulated financial institutions. While bank account penetration is nearly universal (89%) in high-income economies, it is less than half this (41%) in developing economies. Instead, those in developing economies must depend on informal mechanisms to protect themselves against risks such as unexpected illness, crop failure, and job loss or make household investments in durable goods, home improvements or school fees. Among other benefits, financial inclusion lessens dependence on rapacious money-lenders who often charge 2% interest per day, equivalent to a ruinous 217% per year.

Until recently, however, financial institutions felt that financial inclusion was a costly exercise with limited potential to generate revenue. Private sector engagement was limited or distorted in these so-called 'thin', underdeveloped markets. Those in India, Pakistan, China, Bangladesh and elsewhere have since discovered that the reality on the ground is altogether different,

---

[84] www.cgap.org

and that, over the medium to long-term, they can benefit from the fortune in small deposits lying at the bottom of the income pyramid, while at the same time discharging their corporate responsibility to society.

Showa Mshanga's story provides a good illustration of this. Showa is a community health entrepreneur in Zambia's eastern district of Chadiza. She is part of a new network of social entrepreneurs who buy items such as medicines, water purification tablets, and household hygiene products to sell them door-to-door in the local community. Having first received business and healthcare training from the government, she is slowly but surely seeing her business grow. Not only has this has allowed her to provide for her elderly father and look after her late sister's orphaned twins, but it has enabled her to underwrite a loan to her uncle so that he can buy new seeds and diversify his crops.

With profits measured in the tens rather than hundreds of dollars, this might not seem much, but it's enough to allow an entire extended family claw their way out of poverty. This and similar types of micro-enterprise across southern Africa are stimulating significant economic growth in areas where as many as two thirds of people live in absolute poverty. As Showa puts it, "I'm building a business while at the same time helping the community and saving people's lives."

The game-changer, and key to her success, is that the small amounts of income she generates are immediately put into an interest-bearing savings account via the solar-powered, satellite-linked ATM recently installed in her village. More than that, "The benefits of the cash machine are not just financial," she says, "but make me feel more secure and part of something larger."

Financial institutions across the board have recognised the opportunity afforded to people like Showa by recent advances in technology – particularly in mobile payments and next-generation ATMs – to cross sell forms of financial products other than savings and loans such as micro-insurance (both life & non-life) and micro pensions, all of which play a part in slowly-but-surely stimulating economic improvement at family, community and national level.

At the macro level, the benefits of financial inclusion are even more pronounced. With social security transfers in the form of pensions, monthly aid to handicapped persons and other state benefits paid directly into individual bank accounts, transaction costs, bureaucratic inefficiency, and the opportunity for corruption are reduced considerably. For the recipient, there is one other great advantage: they don't have to waste time and bus fares on travelling for miles to collect their benefits in person.

The ATM is crucial to all this. In India, for example, the Post Office together with 21 government-owned banks have installed over 50,000 new ATMs since 2014 under its *Aapka Paisa Aapke Haath* (Money in Your Hands) anti-poverty financial inclusion programme, the aim of which is to transmit social welfare payments direct to beneficiaries' debit cards. This has not only introduced efficiencies by streamlining the previous – and much corrupted – system but is stimulating social transformation through the empowerment of women across vast swathes of rural India who previously found it threatening to enter a bank branch. One lower caste woman in Utar Pradesh put it like this, "The ATM does not insult us, or make us feel small." The implications for a country riven by caste are self-evident.

Part of the logic of the government's programme – and it's the same with neighbouring Pakistan and Bangladesh as well as further afield in Indonesia, Vietnam, the Philippines and across the Pacific into South America – is related to natural disasters. All these countries are threatened every year by earthquakes, volcanic eruptions, floods and hurricanes (typhoons). As was discovered during the Typhoon Haiyan disaster response in the Philippines and, more recently, in Haiti following the passage of Hurricane Mathew, the sooner disaster survivors have access to physical currency in the wake of natural disaster, the better. In flood-prone areas, ATMs are placed on top of raised platforms, while in coastal areas at risk of tsunami, they are sometimes even protected by specially built brick walls.

Disaster preparedness and response is exceptionally important in the context of financial inclusion. Countries with higher concentrations of poverty, weak infrastructure, and poor public services are more at risk from disaster shocks. And it's not just the people that are affected; experience has shown that financial institutions serving at-risk populations are as vulnerable as their clients to natural disasters and political crises. While the initial humanitarian and emergency response to crisis remains crucial, there is a growing recognition of the value of disaster risk reduction (DRR) strategies in preparing for and thus reducing economic losses associated with disasters.

Disaster risk reduction programmes improve the resilience of financial service providers serving vulnerable populations as well as the communities they serve. This seems self-evident. But a general consensus among key financial institutions around effective disaster risk reduction practices is still in its early days and more effort is needed to strengthen the capacities of financial service providers and their clients to anticipate, cope, and recover from the negative impacts of disasters.

ATMs play a key role in stimulating post-crisis recovery.

Given the commonly held assumption that advances in financial technology would herald a more or less overnight move away from cash-based economies to ones where cards, biometrics and mobile wallets would predominate, the centrality of the ATM's role in reducing poverty and building resilience to disaster might appear somewhat counterintuitive at first.

Having first acknowledged that the provision of affordable financial services is one of the key drivers behind reducing poverty in both developing and developed countries, we also have to acknowledge that technology has lowered transaction costs to the point that bringing innovative financial services to the world's unbanked and under-banked is now feasible.

Establishing and running bank branches, especially in isolated rural communities, hasn't got any less expensive in the meantime. So, with the ATM channel estimated to be around 75% cheaper per transaction than at a bank branch; next-generation ATM's now having sophisticated cash management capability; the cost of installation and maintenance falling; and customer reach growing, aid agencies and financial institutions alike have little choice but to consider ATMs as central to their overall financial inclusion strategy.

Governments like it too. According to a study published in 2016 by the UN-facilitated *Better Than Cash Alliance*, this is because digital payments have the potential to boost tax revenue and drive economic modernisation. In Tanzania, digitisation is reported to have lowered the cost to government of managing money by 40%, reduced 'leakage' through the payment of bribes to corrupt officials by up to 80%, and increased tax revenues by 20%. It has also had a positive effect on people's trust in government by increasing transparency.

Financial technology has enabled this change. ATMs do more than burp cash and answer balance enquiries; modern ATMs take deposits, recycle banknotes, and make transfers, too. In the process, video links foster financial literacy. Having started the 'cash revolution' fifty years ago, ATMs are now back in the thick of the action.

In terms of financial inclusion, governments can reap efficiencies by using ATMs to facilitate government-to-person (G2P) payments and registered customers can use ATMs for person-to-person (P2P) services including, crucially, distribution of remittances from abroad. A person in the UK or US wishing to send remittances to family members in rural India or in the urban slums of Bangladesh, for example, can use a service that will allow the unbanked recipient to access the funds without a bank card at an ATM via an SMS message or QR code sent to their mobile device. This not only avoids the recipient having to make long journeys and wait for the nearest remittance agent, but reduces transaction costs, reduces the potential for fraud, and generates new revenue streams for ATM deployers.

The other thing to realise is that ATMs don't replace but complement the physical and digital channels employed by financial institutions. It's not a question of mobile *or* cash, but a question of mobile *and* cash. The clear intersection between physical and digital channels enables ATMs to significantly drive down costs in the payments space for financial institutions. And that's not all.

The introduction of so-called 'thin' client solutions in mid-2015 has transformed the ATM channel. According to the tech company NCR, this has "sparked the biggest change to the way an ATM operates since the self-service banking channel was invented nearly 50 years ago."

Traditionally, ATMs have run on PC-based 'thick' client software and hardware technologies physically embedded in the machine. Now, new state-of-the-art hardware has been combined with cloud-based software in a way that not only makes transacting faster, safer, and more seamless, but has reduced the total cost of ownership by up to 40% while doing so. For a bank or independent operator running a network of 100 ATMs, thin solutions like this could conceivably reduce the total cost of ownership by $540,000 to $800,000 a year by breaking free from the constraints, expenses and security procedures of existing ATM technology.

Cost reductions on this scale mean that the total ATM footprint can extend further, especially into urban slums and remote rural areas that were previously thought to be too difficult – i.e expensive – to reach.

There are lots of things to consider as more people have access to financial services, such as the impact on the ATM channel and cash management. Generally, moves to improve financial inclusion are seen as pushing consumers towards greater use of electronic payments. This is certainly true, but rising numbers of account holders also means more people using ATMs and therefore more cash in the system. For instance, the World Bank says 230 million people in South Asia with an account pay utilities or school fees in cash. It's just one example, but it highlights that financial inclusion does not just mean a move from notes and coins to debit cards and mobile money. This has implications at the operational level for the cash management industry as well as at the national level for central bank monetary policy.

The burden does not have to be borne by the banks alone, however. Financial inclusion strategies can include the use of white-label or brown-label[85] ATMs to bring more players into the market and help bring down costs further. Regardless of the deployment model, there are clear advantages to deploying

---

[85] A 'white label' ATM is installed, owned, and operated by a non-bank deployer. A 'brown label' ATM is where the hardware and lease of the machine is owned by a service provider, but cash management and connectivity to banking networks is provided by a sponsor bank whose brand is used on the ATM.

ATMs at the centre of any financial inclusion strategy from both a customer and financial institution perspective.

Demonetisation is not accelerating financial inclusion

To those critics with an interest in seeing the end of physical currency, cash is a pointless relic. It is, they argue, highly inefficient, being expensive to produce, process, and protect. Even worse, it allegedly props up the shadow economy due to the anonymity it provides for illicit transactions. As a direct consequence, governments around the world are increasingly placing restrictions on its use. They cite that such restrictions increase accountability, make good governance more transparent, and helps in their battle against terrorism, money laundering, tax evasion, drug dealing, and the criminal underworld in general. On top of that, at the macro-economic level too much cash in circulation prevents central banks from setting monetary policies that stimulate economic growth. Apparently, too much of our cash is being hoarded under our mattresses is therefore not being used to bolster production or stimulate consumption. This is putting a brake on the world's economy. The sooner we get rid of the stuff, the better.

The trouble is, most of us haven't read the memo. As with the printed book and vinyl records, it turns out that cash has not been rendered extinct by the smartphone at all. In fact, people – including millennial's – like cash more than ever. Cash is still overwhelmingly the dominant method of payment around the world.

Strange, then, that on 8 November 2016, India's Prime Minister, Narendra Modi, followed the European Union's decision to stop printing the 500 Euro note by announcing he was taking high-denomination banknotes out of circulation, effectively rendering 86% of the country's currency illegal more or less overnight. In a single stroke, this ill-advised attempt at social engineering eroded the livelihoods and savings of almost half a billion people. Given that 95% of all transactions in India are made with cash and where an estimated 45% of the economy is informal, this was bound to have serious unintended consequences.

The most obvious of these consequences was that the 'black market' moved into gold instead. "With a whole bureaucracy geared to ensuring legal importation," said one commodities trader in Mumbai, "all that's happened is that smuggling has increased by unbelievable proportions." If the country's demonetisation programme was intended to coerce commercial and political interests away from the shadows of the grey economy into the sunlight of political and economic transparency, it had the opposite effect.

As if further rebuttal of this policy ere needed, Forbes magazine's editor-in-chief Steve Forbes dismissed this overnight move as "the most extreme and destructive example yet of the anti-cash fad currently sweeping government around the world." The Wall Street Journal also slammed the banknote ban. In an editorial, the newspaper pointed out flaws in the implementation of what it termed India's "bizarre war on cash" and referred to a guest column written by Amit Varma in the Sunday Times of India where he called the move, "A humanitarian disaster." Varma highlighted how the poorest in Indian society – the farmers, landless labourers, and daily wage workers – had been rendered helpless by the move, and how they had had to resort to begging and bartering to survive as liquidity evaporated and ATMs ran out of money.

As this demonetisation programme unravelled into a spiralling disaster, the government began to look for a new narrative to justify the policy, eventually settling on the argument that demonetisation would enable a quicker transition to the cashless society they were seeking. But, as in other countries seeking such 'cashless' solutions, this argument ignored a number of inconvenient truths.

First, a cashless society is not necessarily in the best interests of the majority, as it tends to favour a few at the expense of the many as those most likely to experience a crisis are the least likely to have access to e-payments technologies. These are usually the poorest in society who end up paying a 'poverty premium' for accessing their digital money.

Experience nevertheless indicates that widespread access to cell phones and electronic money transfer systems permitted a major response to the Somalia famine, the Syria crisis and other recent humanitarian emergencies. But cell phone networks do not exist in places such as South Sudan or the Central African Republic and can be shut down by governments (as they were during the Lebanon crisis). Money transfer companies have also been targeted in the 'global war on terror', underlining that these technologies are neither risk-free nor guaranteed, even where they work well[86].

One of the groups that clearly stands to benefit from cashless transacting is the financial sector. Whether payments are made using mobile wallets, debit cards, or wire transfers, there are middlemen involved in every transaction. The fees charged for this intermediation results in huge gains in revenue for everyone involved, including the banking, credit card, telecoms, and payments companies. This additional revenue has to come from somewhere ... and of course it comes from the pocket of the average consumer.

---

[86] ODI and Tufts University report: Is the humanitarian sector fit for purpose, November 2016

And while it is true that transaction fees tend to drop as more people use electronic banking, they will never be zero; there will always be a social price to pay when going cashless. This partly explains why the ATM industry Association holds the view that a cashless society would almost certainly deepen already fragile social divisions between rich and poor, have and have-nots, between the banked and unbanked, and along the digital divide. "The abolition of cash," they say, "could seriously endanger social cohesion and set back any upward mobility between the informal and formal economy.

There is also the real risk that any shift of resources and power to the financial sector only increases income inequality. High volume and low margin deposits from 'the bottom billion' generated through top-down financial inclusion strategies will enable more lending but this is unlikely to be available where it is needed most – to the poorest in society – as the risks involved are still considered to be too high, and the returns too low.

Second, the other main beneficiary of a cashless society will be 'Big Government'. One of the most likely unintended consequences of going cashless will be an increase in bureaucracy in the name of 'accountability'. While lauded as a necessary part of the so-called "democratisation of money" process, what this will mean in practice is further strangulation by red tape that is certain to impose ever-more complicated labour, licensing, taxation and other regulations, which, while increasing tax revenues in the short-term, will end up stifling entrepreneurship and innovation over the longer term.

Furthermore, by controlling people's wealth through the banking system, governments will have complete control over taxation. In the short term, this will be no bad thing. As has already been mentioned, there is already considerable evidence to show that tax revenues climb steadily when digital payments become the norm. But this will come at a longer-term cost, as it will become easier for governments to impose and collect additional taxes whenever they want. With their hands perpetually in the till, governments could be tempted to tax and spend at will. In a country like Sweden, where government is trusted, the legal system fair, and the infrastructure solid, this poses little threat. But the same cannot be said of a country like India ... or Mexico, or Venezuela, or Nigeria, or Rwanda, all of whom have stated their intention to go cashless as soon as they can. Once the state has total control over its citizens in a cashless society, anonymity would be lost.

At the same time, if a rogue government wanted to penalise an individual or, worse, a minority group whose views they don't agree with – think Turkey and the Kurds or Myanmar and the Rohingya – it could do so very easily by freezing their access to digital money. That the state will have this power is the real risk of going cashless.

Third, there is a growing sense that the lack of privacy connected with use of electronic money hands too much information, and therefore too much power, to governments regulators and corporate enterprises. With digital money, all records of what you buy, where, and when can be accessed legally by the government and by direct marketers who want to sell you things you don't need and can ill afford. Of course, the same information can be illegally hacked by anybody with the requisite skills and used against you. The anonymity that comes with using cash will be destroyed, as will people's ability to keep their finances and their purchasing history private. Cash empowers its users to buy and sell what they want without fear of surveillance.

Cash promotes liberty by allowing us anonymity. This is why governments hate it. Governments hate money because they find it difficult to control. If we are always paying electronically, governments always know where we are and what we're doing at any given time. Marketeers build up our psychometric profile to shove yet more targeted advertising our way. This is just another step in the ever-growing power of government over our lives, and the manipulation of our wallets by greedy consumer moguls.

Fourth, cash is universal and final. It is accepted by nearly all businesses around the world, especially small businesses for whom a daily positive cash-flow is essential to survival. Cash offers small businesses instant, free, and guaranteed transaction whereas digital money does not. Cash payments are also 'final' in as much that payments cannot be reversed unless by mutual agreement. Going cashless demands a reliable, universal, and fully functional e-payments infrastructure. This doesn't exist, even in high income societies, and, barring the odd Nordic country, probably never will.

Fifth, relatively speaking, cash is extremely secure. Of course, the threat of armed robbery is not very pleasant to consider, but it has to be measured against the online insecurity that comes with paying electronically. The risk of having your money stolen through cyber-fraud is thousands, if not hundreds of thousands, of times greater than the risk of being mugged … at least, in monetary terms. Compared with the scale of cyber-fraud, robberies involving cash are peanuts. Given this, it can be argued that, far from being a liability, cash provides a bulwark against cyber-crime.

And finally, cash empowers people … especially poor people. Ask the young wife in a Mumbai slum who hides spare cash from her alcoholic husband, or the old mother in Moscow who stuffs banknotes under her mattress because it gives her a sense of autonomy. Ask Showa, the small business entrepreneur in

Zambia who feels that knowing how to use a cash recycling ATM makes her connected to the future.

All this begs a few questions: Why, when at least half the world still prefers to use cash when making small purchases; when the number of ATMs around the world is growing at the rate of one every three minutes; when cyber-fraud is becoming a serious threat; when cash remains the only viable means of exchange in times of crisis, when systems are down and electricity in short supply, are governments intent on doing away with cash? Isn't removing physical currency from our pockets and purses a self-defeating act of economic sabotage?

As David Butterfield, a fellow in Classics at Queen's College, Cambridge, wrote in the Spectator on 28 January 2017, "We're told that digital payment is a welcome liberation from the shackles of cash. But exclusively digital payments actually restrict the reach of money. Cash is a sober reminder of what money stands for. It promotes independence and engagement. By contrast, a cashless society is soulless and joyless."

When views like this, and the 'inconvenient truths' behind them are factored into the mix, it becomes clear that the imposition of a cashless society would be inequitable as well as being economically and socially backward.

Cash is more than important; it is vital, for it keeps our options open. We should worry about the imminent demise of cash because physical currency possesses worth far above its face value. Abolishing physical currency in favour of its digital equivalent is antithetical not just to economic liberty and freedom of (payment) choice, but to the core social values of privacy, protection and empowerment. Worse is the lasting damage to the fabric of society which is likely to be brought on by limiting access to cash.

ATMs have also played a key role in other forms of crisis situation. Take the case of Libya, where, following the overthrow of long-time ruler Colonel Ghaddafi in 2011, the incoming authorities regarded replacing his portrait on the country's banknotes as an economic and political necessity. This, they argued, was to signal "a new dawn" for a country facing disintegrating tribal loyalties and all-out civil war.

Banknotes are more than just glorified vouchers of nominal worth. Often exhibiting intricate historical, archeological or natural designs, they reflect the cultural values of the country they represent. With the ruler of the day usually prominent as well, they also reflect the political mood of the time. This is to say that banknotes play an important – but often overlooked – psychological role in fostering national identity.

For a politician, it does no harm to remind the populace of who is – or, rather, who is no longer – running the country, either. Such are the sensitivities surrounding this that one of the first actions of a new government is to remove all symbols of the previous administration and its leadership.

The changes were not immediate, however, given that Libya was ruled by a Transitional National Council for most of the year following the revolution's end, with the newly elected General National Congress not taking power until August 2012. Within less than three months, however, the first of a new series of notes printed by British security printer, De La Rue, could be seen on the streets of Tripoli and Benghazi.

In this politically symbolic move, the first in the new series, the 1 dinar note, was launched on 17 February 2013, the second anniversary of the revolution's beginning, replacing the version that featured an image of the ousted ruler. The changes involved removing all other imagery associated with the Ghaddafi era, too, including the word 'Jamahiriya' – being Ghaddafi's term for his own brand of apolitical democracy – and the inclusion of English in addition to Arabic text. The front of the new note depicted a cheering crowd waving revolutionary flags, while the back featured the Libyan flag and doves. The latest security devices such as the windowed motion security thread were also used for the first time on African soil.

In another of De La Rue's notes, Ghaddafi's image on the front of the previous version was replaced with one of the Benghazi lighthouse ... this being particularly significant for me as it was the place where, under the horrified eyes of my SAS close protection team, I once managed to sneak a quick swim in the Mediterranean between meetings while working for the British 'Embassy' at the height of the revolution in July 2011.

Before any of this took place, however, a worsening humanitarian crisis in the summer of 2011 – exacerbated partly because of the lack of cash – meant the involvement of De La Rue in a slightly different way:

Prior to the uprising of 2011, De La Rue held the contract for printing Liya's currency. But UN sanctions were imposed as soon as the uprising took hold, and pallets of newly printed Libyan banknotes had to be impounded at De La Rue's printing works at Debden in the UK. In August, some six months after the revolution kicked off in earnest, the NTC successfully bid to have this frozen money released to the now largely defunct Benghazi branch of the Central Bank of Libya "for humanitarian purposes". As a result, two hundred million banknotes were flown back to Libya in an RAF cargo plane in the early hours of one hot early August morning. As with all deliveries of large amounts of cash, especially deliveries into conflict zones, the whole operation was kept entirely secret. Even I, as the British government's humanitarian adviser on the ground at the time, was kept in the dark until hours before the cash arrived. I drove out to the airport after midnight with the Ambassador to meet the shipment. The plane flew in to Benghazi's barely functional regional airport in the dark without navigation lights. The ramp was lowering before the aircraft had even taxied to a halt. While one of the four engines was kept running in case of the need for a quick getaway, shrink-wrapped pallets of cash were quickly off-loaded under the protection of armed fighters from one of the pro-government militias who could be dimly seen, weapons at the ready, silhouetted at the edge of the arc lights' radius less than a wingtip away.

The challenges in mounting such a secretive operation are enormous. Not only are there legal liability issues connected with making sure the money ends up with the right people, but in this case it had to be driven without armoured CiT vehicles across more than ten kilometres of contested 'bandit country' where multiple trigger-happy armed factions – each sporting fleets of Toyota Landcruisers to which had been bolted multi-barrel anti-aircraft guns – were wrestling for control. If one word of this operation had leaked to any of these vicious gangs, the convoy would never have reached its destination in downtown Benghazi. In the event, the money reached the vaults of the central bank branch without incident … by which time the blacked-out planes were well on their way back to Cyprus to refuel.

A few days later this money was loaded into Benghazi's 85 or so ATMs. These had been sitting idle without either electricity or cash for months, so had to be serviced at night so as not to alert the population to the fact that cash would soon be available once again. Since the city was under a rigorously enforced nighttime curfew, this didn't pose much of a problem. But the timing of the replenishment did.

Availability of currency is, in itself, a political as well as an economic act. In this case, there was also a humanitarian component as there was little money circulating with which to buy the few commodities then available in the shops. Introduction of limited batches of newly printed banknotes into circulation, even if they did have the hated dictator's face still on them, would send a strong signal to rival political groups that the NTC was indeed in control. If this signal could be sent on the very same day that NATO in the form of the UK's Prime Minister, David Cameron, and France's President, Nicholas Sarkozy were to announce 'victory' in person in Benghazi's main square, then the signal would be amplified not just across Libya but across the world. Another laurel for the ATM.

It doesn't stop there. A series of ATMs once foiled a kidnap attempt in Cambodia. In early 2010, a gang kidnapped the daughter of a wealthy businessman and demanded a ransom. They demanded that payments be made via a series of eight ATMs, with the location of each only revealed after successful release of funds from the one before. With only minutes to react, the police always arrived after the gang had fled the scene. But by monitoring the sequence and locations of ATMs being used, they began to see a pattern emerging. Just before the eighth and final attempt, they predicted which ATM the gang would use and staked it out. It worked. Not only was the gang arrested – and later convicted, partly on the basis of photographic evidence from the ATM's cameras, including the number plate of the motorcycles they were using – but the hostage was found unharmed. The father got his money back, too.

Closer to home, worries about the role money plays in disease transmission crop up on a fairly continual basis[87]. Are such concerns justified? Brand new banknotes, even the newer polymer ones, may appear crisp and clean but can they be home to more germs than a household toilet, as some have suggested? According to Yves Thomas, chief medical researcher at Geneva's University Hospital in Switzerland, they are. This, she says, "is largely because viruses and bacteria which normally live for only a few hours on dry open surfaces such as doorknobs can survive on relatively warm and moist banknotes for up to 17 days waiting to be transmitted to the next unsuspecting human."

It's hard to think of a physical object that is handled more often by a more diverse range of people than a banknote. With between 20 and 100 million banknotes changing hands in Switzerland alone each day, she suggests that an outbreak of, say, Bird Flu (Avian Influenza) could be accelerated and prolonged thanks to the millions of banknotes in circulation. She went on to say that "this unexpected resilience suggests that this sort of inert, non-biological support should not be overlooked in pandemic planning."

The Ebola crisis shows that she may have a point. During the early stages of the 2014 Ebola epidemic in West Africa, an urban myth suddenly sprang up saying that paper money was responsible for spreading the disease. Almost overnight, markets closed and it became impossible to purchase anything with cash. I was at the epicentre of the outbreak in Kenema Hospital, Sierra Leone at the time and witnessed the overnight return to a barter economy for myself. Amongst other aspects, it meant I couldn't

---

[87] The New York Times ran an article about the number of bugs found on ATM PINpads as recently as 22 November 2016.

pay my guest-house bill ... even when wearing rubber dish-washing gloves.

But is this an urban myth or could it be true? As Dr. Thomas already pointed out, it is technically possible for bacteria and viruses to be transmitted from hand-to-hand indirectly by paper money. But, in practice, those infected with the Ebola virus had to display symptoms before the virus could be transmitted, and the likelihood of anybody ill enough to be displaying symptoms wanting to touch money, let alone being in a state where they were capable of spending it, was extremely remote. Nevertheless, it took many weeks before the scared citizens of Sierra Leone felt confident enough to begin handling money again.

But what about the role of ATMs in disease transmission? Could ATMs, especially cash recycling ATMs, accelerate the transmission of communicable diseases such as 'swine flu' in an outbreak? Since there are only three in the whole of Sierra Leone, and only one of them works at any given moment, the Ebola story is not in a position to provide much insight. However, one American story might be instructive: Studies in the US have shown that the vast majority – over 90% – of US dollar bills are contaminated with cocaine. Since the 'vast majority' of US citizens are hardly likely to be 'coke' addicts, how can this be? Apparently, drug traffickers use cocaine-sullied hands when counting out their illicit gains, and many users roll bills into sniffing straws. The residue from such activities is enough to contaminate the rollers and brushes in ATMs, which then go on to contaminate the rest of the money supply. Wherever cocaine can go, germs are sure to follow.

The ATM has created more jobs than it has destroyed.

When the ATM first appeared, bank staff were concerned – even on the day of its unveiling – that this cash dispensing robot was going to make them all redundant; they feared that mass unemployment would be the inevitable result of the 24/7 efficiency that an automation revolution was about to be unleashed. After all, the ATM was the first visible sign to the population at large that there even was an automation revolution. This fear has a long history, first emerging back in 1839 at the beginning of the 'industrial revolution' when Thomas Carlyle railed against the potentially destructive power of mechanisation by warning that "the substitution of machinery for human labour may render the population redundant." But in the modern era, panic attacks about 'technological unemployment' first really struck the high street in the mid-60's with the advent of the ATM.

But was such anxiety really justified? Certainly, had ATMs caused widespread unemployment, the social impact would have been profound. But evidence from the past 50 years shows that this anxiety was – and is – unfounded. In fact, the ATM has ultimately created more jobs in the banking sector than it has destroyed. In a speech to the US Congress in 2014, President Obama stated that, "there are even more bank tellers employed in the US today per capita than there were when the ATM was invented."

This is partly explained by the fact that replacing a few bank tellers with an ATM makes it cheaper to open new branches, creating many more new jobs in sales and customer services in the process. In less developed parts of the world, the relative sophistication of today's ATM has also enabled the rise of the so-called 'branchless bank' which is serving to accelerate 'financial inclusion' among the world's un-banked. This might not have expanded the number of branches, but it has certainly expanded the number of touch-points where consumers come face-to-face with banking services. ATMs do not so much replace bank tellers, then, as require them to gain new skills to complement them. This

227

has meant an explosion in job opportunities for everyone involved in the cash management cycle. That this is set to change, and change very quickly, seems immutable. According to Erik Brynjolfsson and Andrew McAfee in their book *The 2nd Machine Age*, half of all secretarial jobs have disappeared since 2000 and one third of all low-skilled, white-collar jobs will disappear across the world by 2035 because advances in technology are allowing for more efficient automation.

Arguably, it was the ATM that ushered in this brave new world. Depending on what type of job you had, this was either a good thing or a bad thing; standing in line at inconvenient times disappeared more or less overnight. But a mere machine, however smart its operating systems, cannot replicate a human's empathy and problem-solving ability. Traditional robots are actually quite boring. They can usually do one thing really well, but as soon as their operating environment changes, even minutely, they can't cope. Despite the fact that all accidents with driver-less cars are caused by human error, it is still only humans that can cope with the unexpected. Knowing this, consumers still want to interface with a human being and be comforted by the fact that a fellow human is assuring quality and providing technical oversight.

According to Steve Sands of tech company Cambridge Consultants, "the ideal scenario (for the robotic world) will be when automation takes over all repetitive, heavy, dangerous, or very precise tasks, leaving their fallible human operators to add value through making use their creative and cognitive skills." He could have been describing the next generation ATM.

What the machine has always been able to do is carry out simple functions accurately and repeatedly for different users. These functions have grown more and more sophisticated over the years, to the point that it is now able to emulate most of the aspects of the human teller it complements by linking customers to remote counterparts in video call-centres far away. This virtual interface is useful and marks another step in the machine's evolution from automated teller to branchless bank. Automation has allowed the cash management industry to come of age.

chapter seven

revolution

: ~' /revəˈlōō/ʃ(ə)n

noun
radical and pervasive repudiation or replacement of an
established social, economic or political structure or
institutional system

THE FUTURE LANDED IN FRONT OF ME with a heavy,
metallic thump. This was more than a little inconvenient. Not
because I wasn't into the future, I was. Like every British eight-
year-old boy back then, the military manoeuvres of my boyhood
hero, 'General Jumbo' and the miniature army he controlled
remotely from a touch-pad on his wrist was the weekly focus of
my *Beano* comic-reading life. I yearned for the day when machines
would do my chores for me and I too could dictate the fate of
nations from my wrist. It wasn't to be.

No, it wasn't the future that bothered me. It was just that it
had landed noisily on my homework. Slightly confused and not a
little annoyed that my academic efforts had been so abruptly
thwarted, I looked at the metal and plastic object in front of me.
Being not much more than a counter, some buttons, and what
appeared to be a rotating lavatory brush, it didn't look much like

the future I must say. But, though none of us around the kitchen table knew it at the time, it was.

What my smiling father had brought home from the office was a prototype of one of the parts so essential to making the 'automated cash dispensing' project he was working on at the time actually work. It was that bit of the machine – the machine we now know as the automated teller machine – that spins around at high speed counting out banknotes. The bit you can still sometimes hear whirring and clunking in ATMs today; the bit they call the 'currency counter'.

Shirtsleeves rolled up, my Dad demonstrated to his rather unenthusiastic children clustered around the Formica-topped kitchen table how the machine worked. We watched as he inserted a stack of fifty £1 notes – a fortune to us – into a slot in the top of the machine. Having plugged it in, he turned it on. The little plastic brushes whirred into life and started to spin. One by one the banknotes were gripped by the brushes and dragged down through the mechanism, with each one then counted on the mechanical counter on the front of the machine. The numbers increased at a speed too fast to count … as the notes began to be unceremoniously ejected at high speed all over the kitchen floor. Fifty notes went in at the top but the counter had registered less than forty of them. It was clear, even to my 5½-year-old younger brother now scrambling about on the floor collecting the fallout from this demonstration, that the design phase had some way to go. It was no surprise, then, that those early iterations of the cash dispenser took another track and chose to deliver pre-counted packets of banknotes instead.

Things have moved on since those days and the way we pay for things has changed more in the past ten years than in the previous one thousand. Technological innovation extending through digital payments, mobile money and crypto-currency exchange has created a post-industrial 'cash revolution', the latest iteration of which is collectively known as *FinTech*[88].

It can be argued that the automation of retail banking in the form of the ATM constituted the original FinTech application. In the same way that the invention of paper money in the 13th century kicked off a 'radical and pervasive repudiation and replacement of the established financial and socio-economic ecosystem', the cash machine heralded that moment in history when automation began to take the drudgery out of our lives ... at least, as far as paying for things was concerned.

This might be a bit of a stretch, especially when one considers that the 'revolution' in financial transacting really began in the late 19th century with the advent of radio telegraphy, which allowed electronic transmission of financial payments information around the world. But certainly that era of early innovation from the late 19th Century until the mid-1980's – the era now known as 'FinTech 0.1' – reflects a time in which banks were the first to embrace the new technologies of the day.

FinTech is a very broad sector with a long history. Most people hear the term and think about the latest mobile app which can help us pay for our morning coffee without ever having to swipe a card or fumble for a banknote. But technology has always played a key role in the financial sector in ways that most people take for granted and don't ever see. It wasn't just radio telegraphy.

The pace of change quickened in the late 1950's after the Second World War when secret war-time code-breaking technology developed by Alan Turing and his team at Bletchley Park – one of whom, Anne Murray, was my Mum's sister – laid the foundation for modern computing. The 1950's brought us credit cards with readable magnetic stripes to ease the burden of carrying cash. The 1960's brought ATMs to replace tellers. In the

---

[88] www.forbes.com/sites/falgunidesai [accessed 14 Feb 2016]

1970's, electronic stock trading began on exchange trading floors. The Internet allowed e-commerce business models to flourish in the 1990's. And mobile money was available on our smartphones by the 2010's.

In his book *The World is Flat*, Tom Friedman reflects on how technology has transformed the way we do business. "It is now possible," he says, "for more people than ever to collaborate and compete in real time with more people on more different kinds of work from more corners of the planet and on a more equal footing than at any previous time in history." This period is notable for being the phase in which financial services moved from an analogue to a digital approach on the back of considerable investment by the banking industry in information technology.

The ATM's evolution from stand-alone cash dispenser to fully networked machine in the late 1970's is one early example of this type of transformative innovation in the application of financial technology. Had the 1960's been an era of entrepreneurial start-up rather than the era of lifetime loyalty to a single employer, it is not too far-fetched to suggest that the story of the ATM would be the sort of archetypal FinTech start-up story we would equate with Silicon Valley today.

Today's FinTech companies offer secure mobile wallets and payment apps which allow unbanked populations to safely store their money and make purchases without having to worry about storing or carrying large amounts of cash around. This type of dynamic allows small FinTech startups to play the role of banks by helping people to borrow, spend and pay using mobile apps and websites, bypassing banks altogether. The shockwaves generated by this new approach are reverberating through the banking industry still. But does this make the retail bank redundant? Will we be able to walk into a high street bank branch in the future? In its 2015 global banking annual review, the consultancy McKinsey warned that commercial banks "face a wipeout" in some areas of financial services, and that most, including the larger ones, are in a "high-stakes struggle" to defend their business models against the digital disruption posed by FinTech.

"Everything about our financial services experience will change," says Taavet Hinrikus, co-founder of the international money transfer company TransferWise, His sentiments were echoed during a 2016 retreat for global leaders at Davos where Lionel Barber, an editor at the Financial Times, said, "Nobody wants to be in banking, everyone wants to be in FinTech".

Questioning or predicting the demise of banks, however, boils down to understanding the role they play in different markets and regions. At a basic retail and consumer level, banks collect deposits and make loans, while also facilitating payments and currency exchange. In most developed countries, the majority of households have at least one bank account which they use as their central hub for receiving their salaries, making payments, and saving money. But this isn't the case everywhere. In under-banked or un-banked regions, primarily in parts of Asia and Africa, many individuals hold their life savings in cash and either don't have access to bank accounts or lack the financial literacy to make use of them.

The relationship of people in their 20's and early 30's with the financial system is very different to that of their parents … it's an electronic one, on their smartphones," said Mark Zandi, chief economist at Moody's Analytics. "That can, and will be very disruptive to the banking system." The major banks have all taken steps to address this new reality. But, while the FinTech insurgents are moving and growing quickly, they have still to overcome big challenges of their own before reshaping the industry. They are still relatively small niche players in the sprawling retail banking business. They are not deposit-taking institutions, where consumer savings are guaranteed – at least, in part – by the government. They also lack the legal and regulatory apparatus that traditional banks have built over many decades.

FinTech is flourishing because so much of the spending by banks has focused on dragging legacy systems into the present day rather than investing in the type of 'risky' innovation associated with FinTech start-ups. Banks counter that such risky innovation is only possible because start-ups aren't bound by the same tight regulations they are.

But the fact is that banks accounts are still where our earnings get deposited. These deposits sit with a bank where they are used to underwrite loans and investments. Banks provide a level of insured safety when deposits are too large to sit idle in mobile wallets. Eventually, as people buy homes, buy cars and pay for college, their financial needs become more complicated and require larger borrowing limits. At some point, consolidating services through a bank account becomes an attractive option. While there is no doubt that consumer-focused FinTech startups are capturing market share through the efficiencies and transaction speeds they offer, it is difficult to imagine a future where banks completely disappear.

The ATM will play a pivotal role in the democratisation of finance.

Whatever evolutionary course FinTech weaves, one thing is unavoidable: we are in the middle of a payments revolution which has a long way to go yet in disrupting how cash is managed in society.

Arguably, the most significant legacy of this revolution is likely to be the democratisation of finance. The nature of the current retail financial services model of banking is fundamentally unfair. The costs of the system and the profits of financial institutions are overwhelmingly accrued from fees and charges that hit the poorest hardest. A large proportion of those making transfers are those to whom the average cost is a huge burden. As FinTech extends opportunity and drives change, the end result will be the extension of financial opportunity to many more people. The fees charged will no longer be disproportionate to the service. And savings and investments will accrue better returns for the people that hold them.

This may partly explain why one of the world's largest ATM manufacturers, Diebold, set out in early 2016 to buy its competitor, Germany's Wincor Nixdorf, in a deal worth over $2 billion.

Many commentators were perplexed by the move when it was first announced, and asked why Diebold was expanding its ATM business at a time when people are increasingly going 'cash-lite' by turning to contactless debit cards and mobile services like Apple Pay and Google Wallet? Andy Mattes, Diebold's CEO, was quick to respond by making the case that today's ATMs are not the 'cash and dash' machines that we're used to seeing. "They are sophisticated, high-tech machines," he argued, "that can do 90% of what a human teller can perform while at the same time connecting the physical and digital worlds of cash." He went on to say that "twice as many dollar bills are in circulation now than two decades ago … and Americans and Europeans alike took more money out of ATMs in 2014 than they did in 2004".

What he is also saying is that the ATM is not only at the forefront of the cash revolution, but, fifty years later, is up there still. Not many machines can make such a claim. Further evidence for this claim comes from Mastercard who admit that cash remains the most commonly used method of payment when looked at from a global perspective despite the disruptive onslaught of mobile banking and digital payments. In a 2016 study into the self-service sector, the Mercator Advisory Group in the US asked respondents whether their use of cash had increased, decreased, or stayed the same within the past year. Cash made a strong showing, with 85% saying that their cash use had either stayed the same (64%) or increased (21%). Only 15% said it had decreased. With cash retaining its important role in consumer payments, and with ATMs becoming the most popular method of obtaining cash, it's not surprising that Mercator also found the ATM to be the most popular self-service banking platform.

Today's ATM did not just lay the foundation for the digital financial revolution, but remains at the forefront of it.

Findings such as this make it clear that today's ATM did not just lay the foundation for the self-service digital revolution but remains at the forefront of it by having morphed into an essential complement to mobile banking. Five of the top-25 retail banks in the US have recently expanded surcharge-free ATM access for their customers in realisation of this, and others are set to follow suit. According to Bill Knoll, executive vice president and managing director of Allpoint, a cash services company, "this is a compelling example of how the ATM channel has become interwoven with digital transformation ..."

The consumer, in other words, has decided that on-demand access to banking services is important, and banks must meet this demand with more ATMs, not less, and with better mobile wallet offerings which integrate seamlessly with this physical channel. For the digitally engaged millennial, the ATM is part of the digital service delivery mix, and cash remains a crucial part of their daily life. As the physical component of the digital banking mode, this means that ATMs have to be considered as 'critical infrastructure'. It also means that, as banks reduce branches in an effort to trim costs and increase productivity, the ATM becomes a more, not less, valuable part in the financial services mix.

Cash is the only payment method truly based on demand. Every other payment tool is promoted by various financial entities with a vested interest in providing their own solution for customers. Cash is even targeted by these providers, who see cash substitution as a way of increasing their share of transaction volumes and – because a fee is charged each time – their revenues.

The fundamental attributes of cash – universality, trust, and anonymity – don't appear to have been challenged by the plethora of new payment instruments now coming on stream, and the world of electronic payments is still largely competing with itself. Mobile wallets, for example, have a tendency to consolidate the use of cards rather than replace cash.

A flurry of new technologies is attempting to transform how we pay for the things we buy. So far, these efforts are failing to inspire enough of us to adopt these new methods. This apparent apathy is driven by three factors: Our struggle to differentiate between similar and overlapping apps; worries about security; and that we simply don't see the point and are happy transacting more or less as we are.

In a survey of 16,000 consumers conducted in 2015 by the consultancy Ovum, over one third (37%) of respondents said they knew what a mobile payment service was, but had no plans to use one. Of those who had no plans to use mobile wallets, almost one third cited security concerns as a reason for not doing so, while another third said using cash or cards was easier.

Merchants, meanwhile, are being bombarded with different ways of accepting payments coming at them from all directions, from banks, telecoms companies, and FinTech start-ups. All are either too fiddly or take too long to complete, and all require the payment of an additional fee when accepting payments from smartphones. None, on their own, are compelling enough at the moment to win the competition over who owns the payments value chain. No wonder processing payments via smartphones has not taken off yet, at least not on the scale that would imply a wide and profound interest on the part of the consumer.

Ultimately, the real reason for the staying power of cash is that there are no backup systems if electronic forms of payment either don't exist or become impossible due to phones being lost, security protocols being breached, or the power supply being interrupted. In addition, at least one half of the world's population either doesn't have the option of using e-money or chooses not to. Not having a bank account, relying on 'envelope budgeting', or wanting to protect our privacy should not act as a barrier to accepting physical cash into the financial ecosystem.

Digital money needs a way to put cash into the system, too. In Kenya, where mobile money is common, people do it through human agents. The cash management industry is working on the assumption that, at least in advanced economies, machines will do it instead. The financial crisis sent ATM innovation lower down the priority list for the big banks, but investment is rebounding as developed and developing countries alike are looking at advanced ATMs as a more cost-effective alternative to the expense of running a bank branch. This is changing what ATMs are expected to do. In Latin America, the typical ATM can conduct between 50 to 75 types of transaction, compared with 25 in the US and 30 in Europe. In South Africa, newer machines have over 200 features, including allowing people to open bank accounts, sanction short-term small loans and, perhaps rather morbidly, even pay their funeral expenses in advance.

The change agent driving these developments is 'cash recycling', where putting money into an ATM is becoming as commonplace as taking money out of it. Soon you will be able to send a QR code or similar to someone in another country via mobile phone and the code can be immediately translated into cash at the ATM. Paper cheques can be cashed immediately. In short, a third-generation ATM can do 95% of what a human bank teller can do. This is a far cry – in fact 199 far cries – from the single function available in 1967 of dispensing a pre-determined wad of banknotes.

What this means for retail banking is that the 'Fourth Industrial Revolution' has finally taken hold. This transformation has been driven in part by a breakdown in trust between banks and their customers following the financial crisis of 2008. Fairly or unfairly, customers are still angry that their taxes were used to bail out institutions they perceived to be unscrupulous, if not criminal enterprises without anyone being held to account. These same tech-savvy customers also see bank charges as excessive and bank because they have to, not because they want to. This sense of disquiet has not evaporated. Today's retail financial products and services still look and feel outdated. Technology has moved the goalposts and consumers expect faster, cheaper, and more convenient choice as a result.

When the largest generation in world history, the millenials, command the greatest spending power – which they soon will – the cash revolution will accelerate. As with other aspects of commercial reality, those who adapt fast enough to cater for such explosive upheaval will reap the rewards. The rest will wither, and few in the growing army of potential customers will mourn their passing. Looking at specific services, on average one fifth of consumers think they will use a FinTech provider for day-to-day banking services such as current and savings accounts, credit cards and in-store payments in five years' time. For those bigger financial decisions such as car purchase, insurance, investment, and mortgages, it is likely that traditional structural advantages of banks will be preferred. In ten year's time, however, the picture is likely to change dramatically: 20% of consumers anticipate they will trust technology providers with all their financial service across the board from credit cards to mortgages. By then, there will be a host of new providers and innovative new services. Some banks will take digital transformation seriously, others will buy their way into the future by absorbing challengers, and some will fall by the wayside.

And what, then, of the ATM? After all, on first sight, the ATM appears to embody the face of banking's 'legacy effect' where tomorrow's ideas are squeezed into yesterday's infrastructure like a badly fitted bowler hat.

If cash disappears from our lives, then it follows that the ATM will too. But will it? Will we wake up one morning and stumble down the street to search in vain for an ATM, only to find a bricked-up hole in the wall? Will we really only be able to pay for anything with our mobile phones?

People have been predicting the end of physical currency for as long as the ATM has been in existence. And, with the rise of debit cards, contactless payments, and mobile money, these apocalyptic messages of doom have only got louder. With PayPal splitting from eBay and Apple launching Apple-pay in 2015, mobile payments started to climb, and are forecast to reach $58.4 billion by the time of the ATM's 50th anniversary in mid-2017. This does not necessarily herald the demise of physical currency but surely lays the cards of a cashless future clearly on the table.

We as consumers accept that such technology has the potential to make life more convenient. We know that fumbling with cash and plastic cards is primitive, wastes time, and opens us up the fraud. But we also know that futuristic technologies, especially in their early days, have design flaws, and we will need to know what these flaws are before surrendering our financial transacting to those we don't understand and which were designed by people whose motives we may not entirely agree with. Biometric authentication might make us feel more secure but the reality is that current finger-print scanners in our smartphones are easily confused.

The short answer is that we should be wary. The people responsible for the levels of encryption required to render mobile money safe – the National Security Agency in the US and GCHQ in the UK, amongst them – say that 100% infallibility cannot be guaranteed. "Oh, it's all right," say the bankers, "if you're hacked, you only stand to lose a maximum of what your mobile wallet contains ... and anyway, we are liable". This example of reverse logic smacks of desperation and seems a rather counter-intuitive way of getting us to trust a new technology.

The argument in favour of going cashless is that electronic transacting provides better accountability. E-money has the potential to be more convenient, cheaper, more transparent, quicker to reconcile, and more secure. "But," says Bjorn Eriksson, a former president of Interpol, "there are all sorts of risks when a society starts to go cashless, too." At the macro level, digital 'currency' is not 'legal tender' and its value is therefore not secured by the state. At micro level, the retailer bears the cost of processing physical currency. With digital, one way or the other, the consumer does. Consumers will quickly realise what retailers have always known: that it is they who will soon be paying the charges on all their transactions, and it is they, not the banks or card companies, that will be liable for any losses. This is, anyway, what the banking crisis of 2008 has taught them.

Furthermore, studies into dissociation effect – a psychological effect where consumers feel less direct ownership of their money when it's virtual – have shown that personal spending rises by anywhere between 20% and 30% when transacting digitally. The resulting over-spend blows a hole in personal budgeting and, according to some retail experts[89], will eventually lead to a personal and institutional credit crisis. This might be good for retailers but spells disaster for the individual consumer if they don't improve their budgeting skills.

Our vulnerability to cyber-fraud is increasing.

Depressing as this sounds, there are more immediate but invisible concerns, notably the rising threat to privacy when 'Big Brother' can track what you spend your digital money on in real time, and where.

Meanwhile, our vulnerability to sophisticated cyber-crime is increasing exponentially. At the moment 2p in every £100 is lost through traditional fraud while 7p in every £100 is lost when using contactless forms of payment.

There are theoretical benefits to a cashless society, it is true. Payments may be more secure, faster, and more convenient … although, at the moment, they are none of those things. Digital transfers will supposedly leave less room for things like drug-trafficking, tax evasion, and financing of terrorism … but, again, there is no evidence of this happening at the moment. Human error will be minimised when point-of-sale transacting no longer requires change to be calculated and handed over.

But, depending on your point of view, the two main potential benefits are a double-edged sword: One edge of this extremely sharp sword is that merchants will be able to use smart

---

[89] www.onlystrategic.com

technology to analyse your spending habits and use this information to offer incentives and personalised shopping information. With advances in biometric recognition technology of the type seen when Tom Cruise wanders through a shopping mall in the film *Minority Report*, the relatively benign targeted online advertising of today will become the hard cross-sell of tomorrow's text message. As the founder of Square, Jack Dorsey, puts it, "It's not about killing cash; it's about being able to account for your entire business, and maximising that business's cost efficiency". This might be understandable from the payment provider's point of view but is potentially chilling stuff for us, the consumer.

The second risk is cyber-crime. And we humans are really, really bad at recognising risks that we can't see or have never experienced. We consistently under-rate the dreadful things that can happen to us, while we consistently over-rate the most unlikely of events. It is still 1,000 times more dangerous to drive a car on Europe's roads than to take an international flight, for example, but we are still more anxious of flying than we are of driving.

"We as human beings don't perceive risk rationally," says Martin Hartley, chief operating officer of a risk management company. "The Paris terrorist attacks made the risk seem far worse by its magnitude, not by its probability. That's the challenge of risk." In other words, it's all about perception – of relative and residual risk – and of an individual's tolerance of what level of risk is acceptable.

According to Neal O'Farrell, founder of the not-for-profit group Identity Theft Council, apathy is the biggest problem to digital protection. "The financial services community has been very good at persuading consumers that zero liability means zero responsibility about identity theft," he said, "and one of our biggest challenges is to get consumers to believe that it will happen, and, when it does, it will be extremely painful."

Meanwhile, the biggest logic in favour of going cash-less is also the simplest: Carrying banknotes around is an invitation to get robbed. But having your smart-phone stolen, your SIM-card swapped, or your online bank account hacked puts you at even greater risk of financial loss ... despite the fact that it means zero risk of getting clubbed over the head or shot. Hacking, identity theft and cyber-theft are a gift to modern, technology-savvy criminals who no longer have to go out and face risk of capture when robbing people in the dead of night but can make millions with a phone and laptop from the comfort of their bedrooms.

While this is going on, the volume of physical currency in circulation around the world continues to rise and, with it, the number of ATMs. This is partly because modern ATMs are genuinely 'next-generation' machines. With the right software, hardware and inter-connectivity the ATM can carry out just about any financial transaction these days ... in more remote areas and at lower cost, too.

There is no reason why the ATM cannot offer a range of bank-related functions beyond cash dispensing to include, for example, cash recycling, foreign exchange, cheque deposit taking, as well as other services like cinema and concert ticket purchases, tax payments, bill payments, and mobile phone top-ups. And it doesn't stop there. In 2016, India's IDBI Bank launched an ATM offering that allowed its customers to purchase government securities. It was a first for the country, and a clear example of how the ATM can help bridge the financial inclusion divide.

Employing the latest innovations in haptic, visual, and biometric technology, the ATM is poised to take on the full raft of mundane tasks previously carried out by their human counterparts, the bank tellers, for the first time. All of the day-today technical tasks that would previously have been carried out at the counter can now be done by the ATM, freeing up time for those staff that remain to add value by providing tailored financial advice. Using remote video links and twin touch-sensitive screens, this can be done remotely too, rendering not just the teller but the branch itself obsolete. If this is not about enabling and empowering the customer, it's difficult to know what is.

What is needed is an evidence-based understanding
of the political economy of cash.

So, is the idea of a 'cashless society' based on a set of false assumptions? To an extent, the answer must be 'Yes'. According to the ATM Industry Association, the evidence points to the fact that there is simply no business case for restricting freedom of choice in payments for millions of consumers and merchants through the eradication of cash. This is especially true when the vast majority of cash transactions are perfectly legitimate and made in good faith. But are they? Some argue that the justification for the abolition of cash is its use in the black economy for tax evasion and other forms of criminal activity. But is this fair? Research is badly needed to inform any debate on the political, social, and economic impacts of such decrees and their potential for causing lasting damage to the fabric of society by limiting access to cash.

Paper money allegedly props up the shadow economy due to the anonymity it provides for illicit transactions. As a direct consequence, governments around the world are increasingly placing restrictions on the use of physical currency. They cite that such restrictions increase accountability and help in their battle against terrorists, money launderers, tax evaders, drug dealers, and the criminal underworld in general.

Apart from resenting the implication that I belong to one or more of these criminal groups just because I wish to have some cash in my pocket, restricting access to the money supply turns out to be economically perverse. It's nonsense in other words. According to Mordechai Fein, a former member of the World Bank and IMF, this is because, "there is absolutely no evidence to show that tax evaders or terrorists are threatened and in retreat after cash use prohibitions have been introduced."

Yet, it appears that only criminal considerations have been taken into account when implementing cash restriction measures. Behind the tightening of regulations is a basic assumption that such restrictions will harm and ultimately defeat criminal activity and render the shadow economy more transparent and governments somehow more accountable. This thinking infringes on cash users' dignity by implicitly labelling them as potential criminals. Furthermore, cash users should be entitled to have the option – the human right, even – of using whichever payment mechanism they choose. Not everyone wants governments or faceless corporations monitoring every aspect of their lives. This does not make them criminal.

There is one other political rationale for converting to a cashless society. In his recent book *The Curse of Cash*, former International Monetary Fund chief economist Kenneth Rogoff, says banknotes are preventing monetary policy from fulfilling its potential for stimulating growth. The reason for this, Rogoff argues is that, in a zero or negative interest rate world, too much cash is lying under our mattresses. Apparently, about one third of the world's banknotes are not circulating in society and are therefore not being used to bolster production or stimulate consumption. In the US, he says, $4,200 in cash is sloshing around outside the banking system for every man, woman and child in the country. Household hoarding like this is putting a brake on the world's economy.

The financial logic is fairly simple: When banks can no longer use interest rates cuts to stimulate the economy, institutional investors will ditch bonds and hoard cash. That may be OK when it's just a few people stuffing cash into the cookie jar, but it risks monetary catastrophe when institutional investors do it. Central bankers, who have little room for manoeuvre, may soon come to the point where they have no option but to set negative interest rates so low that a stampede for high-denomination banknotes is triggered.

Limiting the risk of this happening by moving away from paper money could make central banks a lot more effective in stimulating the economy and fighting the next recession.

The abolition of cash could endanger social cohesion.

But there is another, hidden, side to the cashless debate which doesn't get talked about much by those who stand to benefit from going digital: Each transaction generates revenue for everyone involved. It is also a 'one-time' transaction, the benefits of which dissipate as soon as the transaction is complete. Transacting in cash, on the other hand, is completely free of cost and exerts an economic leverage effect as it circulates through local markets.

Meanwhile, we have seen that cyber-crime has become a huge business. Some estimates put the cost of cybercrime to the global economy at more than $445 billion. Compared with the scale of card data fraud, crimes involving cash can be counted as peanuts. Given all this, it can be argued that, far from being a liability, cash provides a bulwark against cyber-crime.

ATMIA goes further by suggesting that a cashless state would almost certainly deepen already fragile social divisions between rich and poor, haves and have-nots, between the banked and unbanked, and along the digital divide. The abolition of cash could seriously endanger social cohesion and set back any upward mobility between the informal sector and formal economy.

The world's appetite for cash is strong and growing.

Even with leapfrogging advances in technology, some aspects of cash and how we use it cannot be reduced to digital bytes. There is simply no alternative payment system that is as reliable, anonymous, or convenient. Cryptocurrencies are anonymous, but are, at the moment, unstable and inconvenient. Credit and debit cards can be skimmed and cloned. And electronic payment systems like PayPal and Apple-Pay require the installation of 'apps' which trace not just your money, your habits, and your location, but into which have been embedded cookies and backdoors to allow for more efficient cross-selling.

Despite the technological advances and the exhortations from those who stand to benefit from a cashless society, a system that combines all the rational and emotional elements that physical cash provides simply does not yet exist. Which is why, when you look at the statistics of how much cash there is in circulation and how it is used around the world, paper money is far from disappearing. The worldwide appetite for cash remains strong.

In the United States, the number of dollar bills in circulation grew 42% between 2007 and 2012 and continues to grow by about 5% year-on-year, outstripping inflation or GDP growth. This is not a byproduct of quantitative easing but because consumers are happy having cash in their pockets and purses. In the UK, the Bank of England reports that growth in banknotes in circulation has outpaced growth in aggregate spending in the economy since the mid-1990s and that just under half of all transactions are still made with cash.

Global ATM cash withdrawal volumes grew by 7 per cent in 2014 with a total of 92 billion withdrawals made, according to Retail Banking Research's (RBR) recent report entitled Global ATM Market and Forecasts to 2020. Growth in the number of ATMs has been even higher, with the global installed base now having passed the three million mark. The value of cash withdrawn from ATMs in the Asia-Pacific region grew by 50% over the five year period 2011-2016, according to more recent research[90] from the same consultancy, with the number of cash withdrawals per ATM also rising sharply. Apart from many of the region's economies being cash intensive, RBR suggests that this increase in demand can be put down to expanding financial inclusion programmes which are bringing large numbers of people into the banking system. Indian ATMs are now used at twice the rate of those in France, averaging 5,000 withdrawals per month. "The runaway growth seen in many Asian markets is set to continue for the foreseeable future," said Robert Chaundy, who led the RBR study, "with the key challenge for banks there being to get enough ATMs installed to deliver cash to their customers."

In light of the so-called 'war on cash' in some countries, these increases may seem surprising. Even countries that are often held up as leaders in the crusade against cash – countries such as Sweden, Denmark, and Nigeria – aren't really intending to get rid of banknotes and coins at all. In the summer of 2015, newspaper headlines around the world declared that Denmark and Nigeria wanted to rid themselves of cash by 2021. But what the finance ministries in both countries were really proposing was that retail outlets should no longer have the obligation of having to accept physical currency. They could go cashless if they wanted. But this

---

[90] Retail Banking Research: 'ATM Hardware, Software and Services 2016' report, 28 October 2016.

will be the retailer's choice, not the consumer's. Consumers will vote with their wallets.

Our psychological attachment to cash; the infrastructure available to banks, retailers and telecoms providers (including the electricity supply); the requirement to create compatible systems; and the regulatory environment, all make the "inevitable progression" away from physical currency riskier than it first appears. Where does that leave us now, then? To paraphrase Mark Twain, 'reports of the imminent demise of cash have been greatly exaggerated'. More pragmatically, "Until we have sufficient and reliable alternatives in place, it would be dumb to get rid of cash now," says David Wolman, author of the book *The End of Money*, adding that, "people and businesses in developed as well as developing countries still rely on it."

He is not alone in having wearied of the digital revolution; of the relentless assault by credit card companies and banks on the virtues of a cashless society. Many of us are privately sceptical that such a thing might be the force for good they say it is. Cash is a global public good and we are suspicious of any motive, real or otherwise, that smacks of privatization; that has the taint of being yet another way to part us from our money while they make money in the process. Making a percentage on every transaction – each of which, unlike with cash, is one-time and one-way – they get inordinately rich. They do so at our expense and make the inequities in society deeper still.

More pragmatically, we are also wary of dark clouds gathering on the horizon that seem to portend of cyber-horrors yet to come. Online fraud has not gone away. Nor has our visceral need for privacy. We should worry about the imminent demise of cash because, as this book has pointed out repeatedly, physical currency possesses worth far above its face value. Cash is more than important; it is vital. For, if nothing else, it keeps our options open.

The ATM is under-appreciated and under-utilised.

For decades now, the ATM has brought freedom, convenience and secure access to our money at any time of day or night. In a 2016 presentation in New Orleans, Mike Lee, a futurist and ATMIA's CEO, offered a glimpse of where he saw ATMs sitting in this faster, more mobile digital world[91]. He forecast that another million ATMs will dot the globe by the end of the decade, raising the total to 4 million and bringing cash access to the remotest of places. He also agreed with the views of Retail Banking Research, that ATM cash withdrawals will grow by 50 percent within that same timespan, increasing from $92 billion in 2014 to $128 billion. In supporting these forecasts, he cited the continuing growth in cash demand over the past decade, up nearly 9% year on year compared with a 2.5% – 3% increase in GDP.

But, despite its appeal and obvious benefits, the ATM is under-utilised. Widely referred to by consumers as "cash machines", the terminals have become synonymous with cash withdrawal. A survey carried out in 2015 in the UK found that almost 96% of those who responded had withdrawn cash from an ATM in the past month, but only 15% had made use of additional services. According to Kirsty Berry of Compass Plus, the payments software company who conducted the survey, "it is the introduction of cash-recycling ATMs, video teller machines, and terminals that allow cheque deposits that will drive reductions in branch footprint."[92] Enabling these features on ATMs, she went on to say, "might also open up opportunities to reach un- or

---

[91] www.atmmarketplace.com/articles/atmia-2016-the-evolving-future-of-the-atm [accessed 10 March 2016]
[92] The UK lost 40% of its total branch network between 1989-2012 and numbers are predicted to fall further – by around 9,800 or 9% by 2020.

underbanked consumers, which would undoubtedly add value to a bank, not only by increasing its reputation and customer base, but also by adding to its bottom line."

Does she have a point? Could ATMs do more? In Russia, the bank MultiCarta lets customers apply for credit cards and other banking products via their ATMs, while Leto Bank, another Russian financial institution, offers customers express loans. In Ukraine, PrivatBank allows foreign exchange transactions and international transfers through Western Union where recipients withdraw remitted funds at an ATM. In China, it is possible to open up a bank account and have a debit card 3-D printed at the ATM in under eight minutes.

Overall, such initiatives suggest that the cost-efficiency and multi-functionality of the ATM is under-appreciated and under-used by both financial institutions and consumers. If used to their full potential, the latest iteration of ATMs can transform themselves into almost entirely self-sufficient branchless banks. ... which is just as well, as, with 40% fewer bank branches than there were a few decades ago, the US, the UK and other high-income countries are in desperate need of ATMs to provide financial services touchpoints in communities around their respective countries where customers are able to deposit cheques, apply for a mortgage, loan or credit card, get a replacement card, carry out a video conference with a mortgage consultant, and more.

As Ron Delnevo, ATMIA's European director, has suggested, "Community 'hub' ATMs capable of offering such a range of services will morph into one-stop personal finance shops for everything to do with financial services in the midst of a revitalised local community." If he's right, this will herald the ATM's second coming.

The ATM once forced us to recalibrate our relationship with money ... and it's about to do it again.

So, 50 years after the arrival of the ATM in our high streets it remains the case that there is more physical currency in circulation than ever before, not less. This trend will doubtless turn as digital comes to scale and trust builds, but it has not yet. "People talk about the demise of physical currency as if it was only a matter of time," says Mike Lee, chairman of the ATM industry Association, "but even 50 or 100 years from now, the printed variety will still be a big chunk of our lives." There are many reasons for industry gurus to hold such views, but before discussing what these might be, a look at the publishing industry might be instructive:

In 2014, one of the world's largest publishers, Penguin Random House, doubled the size of the warehousing operation needed for global distribution of its printed books. Why would it do this in the face of soaring e-books sales? After all, e-books sales were up 1,260% between 2008 and 2010. Bookstores were closing by the dozen and publishers and authors feared that cheaper e-books would cannibalise their business. "E-books were this rocket ship going straight up," said Len Vlahos, a former executive director of the Book Industry Study Group, a non-profit research group that tracks the publishing industry. "Just about everybody you talk to said that publishing was going the way of digital music."

But the digital apocalypse never arrived. Where analysts once predicted that e-books would overtake the print version by 2015, digital sales instead slowed sharply, falling by 10% in the first half of 2015 according to the Association of American publishers. Early adopters of e-books were returning to print, or at least juggling their reading devices and paper. The decline then reached a plateau where it remains today. This decline in popularity of the e-book signalled that the publishing world, while far from immune to technological upheaval, was able to weather the tidal wave of digital technology. It is likely to be like this with paper and digital money. There are already signs that the manic rush to mobile money is resulting in different forms of e-payment taking market share off one another rather than eroding the use of physical currency. This means that, while electronic transacting will no doubt eventually predominate, banknotes will not disappear from our lives any time soon. And, if the example of the e-book market is anything to go by, they never will.

But cash has its digital rivals, from 'chip and PIN' plastic cards, mobile phone apps, contactless cards using 'near field communications' (NFC) technology, cloud-based systems, and even virtual currencies like 'Bitcoins'.

None of these rival systems are perfect, though, although the payments industry is racing round the clock to make them so. To paraphrase *Star Trek*, they also go where no money has gone before. Social media and the internet have democratised what we know about money and how it is managed in a way that simply did not exist a few years ago. In addition, choosing to pay for some instant gratification – a bar of chocolate, say, or a cup of coffee – with the 'digital wallet' lurking in our smartphones separates us from the traditional financial services infrastructure we were familiar with only yesterday. As part of this 'paradigm shift', we are all being forced to recalibrate the way we think about our financial relationships, not just with cold, hard cash, but with those that provide it, our banks.

It was exactly the same when the ATM first burst onto the scene fifty years ago. Its arrival forced us to recalibrate our relationship with money.

Next generation ATMs will recognise the user not the card.

It is likely, then, at least in the short term, that the way we pay for goods and services will involve a mish-mash of old and new. The next generation ATM will be expected to recognise the user rather than the card – in fact, the card will become redundant – and will do this through a combination of behavioural algorithm and physical biometric recognition. But this will only happen if designers create a seamless transition from the real to the virtual, from the tactile to the visual, from the rational to the emotional, and back again.

Spike Jonze poses a teasing question in the 2014 movie *Her*: Is it possible to fall in love with a computer operating system; to become embroiled in an affair with a machine with nothing but an appealing voice; a virtual person which astonishingly seems to have the gift of human intuition and empathy? This is what happens to Joaquin Phoenix who plays a lonely man plunged into melancholy by the break-up of his marriage, invests in a computer programme that promises a virtual companion, finds himself talking regularly to 'Samantha' (huskily voiced by Scarlett Johansson), who somehow seems to understand him and his needs better than his former wife did.

This is where the futurists would like to take us; to get us to believe that we can empathise with robots. But the ATM is no robot, and it still has an enormous role to play in the future development of self-service banking. How the machine interfaces with its human operator will determine the speed and extent to which this is achieved. Human behaviour can be fickle and counterintuitive. If you tell people in California there is going to be a tsunami, some of them will go down to the beach to watch.

There is some way to go in making the ATM truly interactive. It's not so much about automating the teller, but humanising the machine. As with many robotic machines, the key to realising its potential lies in making the ATM human and haptic[93]. This is not as easy as it sounds in a field dominated by the logic and functionality of engineering. The machine we have today is what happens when you design an electro-mechanical device from the components up rather than from the user down. The next-generation machines will be totally different.

Customers are loyal to experiences, not companies. Understanding the customer at the individual and personal level means providing them with an experience that gives them what they want, engages them in a relevant and convenient way, and does so consistently and accurately. Knowing who is standing in front of the machine, what language they speak, and their habitual preferences is just the start.

> *ATM users are caught between two*
> *very different ways of perceiving the world.*

Because the present is always a period of painful change, every generation views the world through the rearview mirror. We spend our lives making what we consider to be reasonable simulations of what was done in a preceding age. We are fighting yesterday's battle. Although, of course, we don't have that image of ourselves, tackling right-brain problems using left-brain techniques is comparable to skiing in a wetsuit.

What is happening at the moment is the changes are occurring so rapidly that the rearview mirror doesn't work anymore. Somehow, we have to shift our perceptive focus through the windscreen in front of us, from the past into the future. Take this book, for example: self-publishing software makes it possible for

---

[93] Haptic devices create the illusion of substance and force within the virtual world by applying touch (tactile) sensation to interactions with computer applications.

every person to become their own publisher. E-books mean we no longer need to mechanically print repetitive text; we can make a book unique to each reader if we want. Culturally, what is happening now is titanic. If we are to meet the future head-on, we need a completely new frame of reference, one where there are many centres and no edges, where nobody is everybody. The key is to understand both left and right brained thinking simultaneously.

Today's users of technology are caught between two very different ways of perceiving the world. On the one hand there is the linear, quantitative world of the banknote and what it represents in terms of privacy, debt, and the "promise to pay"; on the other, the holistic qualitative world of the mobile wallet where value, being in the eye of the beholder, fluctuates. One is left-brained, the other right. These two opposing world views are busy slamming into each other at the speed of electronic light, and the ATM is caught slap bang in the middle.

Marshall McLuhan, the philosopher and communications theorist best known for coining the expression "the medium is the message" and for predicting the World Wide Web almost thirty years before it was invented, understood this. Not only did he predict the internet, but he coined and popularised the term "surfing" to refer to rapid, irregular and multidirectional movement through a heterogeneous body of documents or knowledge. And he did this at a time when the ATM had yet to be invented[94].

McLuhan understood the context of 'the medium' in a broad sense. He identified the light bulb as a clear demonstration of the concept of "the medium is the message". A light bulb does not have content in the way that a newspaper has articles or a television has programmes, yet it is a medium that has a social

---

[94] McLuhan: Understanding Media; Routledge & Kegan Paul, 1964

effect; that is, a light bulb enables people to create spaces during night-time that would otherwise be enveloped by darkness. In describing the light bulb as a medium without any content, McLuhan states that "a light bulb creates an environment by its mere presence." In many respects, the ATM does the same; its mere presence on the street corner creates an environment of choice and the possibility of instant gratification. Cash, the reason for the ATM's existence in the first place, is merely 'content' in this sense.

People tend to focus on the obvious – the content – to provide us valuable information, but in the process, we largely miss the structural changes in our affairs that are introduced subtly, or over long periods of time. As society's values, norms, and ways of doing things change because of tectonic or incremental shifts in what the technology offers us, it is then we realise the social implications of the medium. In the case of cash, these range from the cultural and historical roles played by cowrie shells and salt, through interplay with conditions of quality assay, to the secondary or tertiary effects in a cascade of electro-mechanical interactions that we don't ever see.

By extension, all technologies have embedded within them their own assumptions about time and space. The message which the technology conveys, it can be argued, can only be understood if the technology, the context in which it was produced, and the environment in which it is used – and which, simultaneously, it effectively creates – are analysed together. Neither is completely intelligible without the other, particularly in relation to the technologies that preceded it. He also believed that an examination of this relationship offers a critical commentary on culture and society.

But let's pause for a moment. Where is this telescoping of technology leading? How will people be affected psychologically if society was, either by statute or voluntarily, to become cashless? Unlike the ATM with its physical distribution of cold hard banknotes, the digital wallet has one prime characteristic: it decentralises the user, just like the laptop and the mobile telephone did before it. Digital money turns its 'owner' into information, embedded in the same matrix of information as the digital 'money 'itself. In one sense, we are all Bitcoins now.

Once placed in relation to the computer where his or her bank account information is held, the user is everywhere at once. But the user is simultaneously nowhere. You are everywhere and so is everybody else using the system. One can appear simultaneously at every ATM, at every point-of-sale terminal anywhere on the planet. This condition poses an almost insuperable problem for the anti-fraud people: how can it be verified that the mechanism of exchange, the card or smart phone, is in the same place at the same time as the bank account owner?

The nature of virtual money is that it has no centre and, at the same time, no margin. Centres exist everywhere. As with Stephen Hawking's theory of Black Holes, money as a concept disappears into a singularity, a vortex of relativity where everything has a cost and nothing has value. Worse is that we risk losing our privacy – and, by extension, our identity – in the process.

The social impact of the telephone might tell us something about the social impact of the ATM. The telephone creates a special form of instantaneous contact ... it shrinks space to nothingness all at once. Apparently invented by Alexander Graham Bell to improve his wife's hearing, the telephone represented an improvement over the aural mechanism of bellowing through a megaphone and the visual mechanism of semaphore, both of which were limited by distance. This is what Marshall McLuhan said about the telephone: "It is designed to cut through interference over long distances, eliminate spatial distance, and increase the speed of the human voice. It uses electricity to do this. However, the very act of using the copper wire as an extension of the human voice produces a peculiar result: it obsolesces the human body as hardware. Pushed as a technical process to its limit, a reversal effect occurs. Everyone becomes involved in what was originally meant to be a private communication. That is, the major social effect of the telephone is to remove the identity of the caller".

I wonder what he would say now about the next generation of ATMs?

epilogue

YOU MAY NOT HAVE HEARD OF Arthur Fry, but you will know what he invented. A quiet and understated man, Fry worked at 3M, an innovative technology company based in Minnesota, USA. Back in the mid-1970s, he had been puzzling about what the company could do with an adhesive one of his colleagues had come up with. Few of his executive colleagues could see the point of a glue that, while it kept its stickiness, did not hold things together very well.

One Sunday morning, while singing in his church choir, he became frustrated at how the paper bookmarks in his hymn book kept fluttering to the floor. In that instant, he suddenly saw what could be done with this weak adhesive ... and the Post-It® note was born.

As with so many creative solutions, though, it wasn't all plain sailing. First he had to overcome scepticism within his own team who wanted to focus on something with more immediate commercial application. Second, it was not at all clear that there would be a demand for such a product. But consumers do not always know what they need; it was only when he sent out free samples in the mail that potential customers realised how useful they could be, and that they wanted them.

There are parallels here with the ATM. While the note-counting and self-service technology already existed in 1965, neither Dad's business partners, his team, nor even his own family could see the point of bolting them together in a new way to do something that nobody (apart from a banking industry who wanted to satisfy their Unions) appeared to want, far less to need.

Yet, frustrated by being unable to access his bank one Saturday fifty years ago, and faced with a weekend of screaming children who needed to be taken to the zoo, my Dad came up with an idea that was to catapult the post-war world into a new relationship with its money. This idea changed the face of banking just as fundamentally as the invention of oxy-acetylene welding changed the industrial revolution. But it did more than that. Being the FinTech invention of its day, it was the first machine to realise our dream of instant gratification and of a 24-7, on-demand way of life. This machine changed society forever.

As with all the best inventions, the idea was simple: a cash dispenser that allowed you to get hold of your money fast and for free, any place, any time, come rain or shine. This trusty 'hole in the wall' became the Cashpoint, Bankomat, and Automated Teller Machine we still recognise and walk or drive past almost every day of our lives.

That said, despite the ATM being one of the world's most intriguing and unsung icons of social advancement, I'll be the first to admit that a book about a machine which appears to have the personality of a kitchen dishwasher does not at first sight seem particularly inspiring, especially in a world of self-service where we take technology for granted. After all, those of us who have to drive to work every day don't think too hard about how the car works, do we? We just want the wretched thing to start first time and get us to our destination and back quickly and safely without getting wet. We just want the convenience. It's much the same with the ATM. We walk or drive past one almost every day, but take its magic so completely for granted that we barely ponder it at all, far less think about how it got there, or what it has done for society.

From its earliest days, the ATM provided the nexus that bound together the potentially disruptive and rather 'Orwellian' ideas of automation, identity, and digitisation. In this, it was the original 'FinTech' solution.

To dismiss such a machine does a disservice both to the machine and to the men and women involved in making sure it continues to play the massive role it does in making our world go round. I feel this particular machine deserves more.

You might say at this point that, being the inventor's son, 'I would wouldn't I.' And, to an extent, you'd be right. But it's more than that; I felt I had to write the ATM's story for three very particular reasons:

Firstly, because my Dad never got around to telling the full story, confusions have arisen over its origins. As a result, competing myths have obscured the contribution made to society by this awesome machine. The occasion of the ATM's 50th anniversary in the summer of 2017 marks an opportunity to put the record straight.

Second, and perhaps more importantly, there are very few people of whom it can be said, "they changed the world." But my Dad was one. But, despite numerous awards for his 'services to the ATM industry' and 'to banking', I never thought he received the recognition he deserved. As Trevor Baylis, inventor of the world-changing wind-up radio, once wryly observed, "The Brits are brilliant at inventing, but appalling in the way they treat inventors." Despite the apparent success of his invention, Baylis ended up largely forgotten. I was determined this would not be my Dad's legacy[95].

And third, being a disaster management consultant in my 'day job', I see the important and growing role being played by the ATM in disaster zones and in reducing poverty around the world ... a role in which financial institutions and the cash management industry by and large have yet to see as a fully integrated part of their social responsibility to the communities they serve.

Engineers and business pioneers like my Dad play a vital role in changing the world, in driving forward economic and social development, and continue to do so by delivering products and processes that enhance quality-of-life while at the same time contributing to the common good.

The Internet, radar, computer, jet engine, light-bulb, television, telephone – and now the ATM – are all world-changing British inventions, and, in my view, my Dad deserves to be up there with Robert Watson-Watt, Alan Turing, Frank Whittle and Tim Berners-Lee[96].

Part of the motivation must stem from the fact that, although I was not yet eight years old at the time, I was there at the beginning. I was there as the ATM evolved from idea to machine, some of it around our kitchen table. I was there when my Mum first said – amidst much mirth, and much to her later annoyance – that she couldn't remember six digits. And I was there when my brother Nick said it didn't matter anyway if the baseball rule of

---

[95] The Scotsman newspaper in the UK seems to agree, calling for my Dad to receive a knighthood in its leading editorial comment of 31 December 2004.
[96] In fact, as with the ATM, all these inventions had multiple 'inventors'. Robert Watson-Watt, for example, is given the credit for inventing radar when actually it was a German, Heinrich Hertz, who discovered in 1888 that radio waves could be detected after bouncing off objects.

"three strikes and you're out" was followed and the card was 'eaten' by the machine without dispensing any money if three incorrect PIN entries were made.

It is a peculiarly British paradox, though, that such a creative country is quite so inept at capitalising on its inventiveness. In part, especially in the post-war years, this might be put down to a stiff-upper-lipped reticence to self-promotion. Even today in the UK, it is regarded as 'bad form' to brag about success. This reticence goes some way – but only some way – to explain why the ATM is widely considered an American invention. Overall, though, however you decide to interpret history, and whatever your understanding of semiotics, the ATM deserves to be known as an invention conceived by the British, even if most of its formative years took place in the United States and elsewhere.

To the industry association involved, ATMIA, the answer to the riddle of the ATM's invention is very clear, and clouded only by minor quibbles over definition: While others can claim to have developed the original concept, to have introduced the magnetic strip on the back of a debit card, or to have networked the machines to their banks for the first time, the person who first came up with a working ATM – a stand-alone and automated machine capable of dispensing cash quickly, conveniently, accurately, and securely outside and out of banking hours – was my Dad, John Shepherd-Barron.

That various patents are held in various names, at various dates, for various different parts of the ATM and its operating systems – none of them by my Dad – is not, and never has been, in dispute. But all these patents, including that for the 'triple use' Personal Identification Number (PIN) system we use today, were acknowledged by the UK's Intellectual Property (Patents) Office in 1966 as having been pre-empted, and therefore subject to legal protection under *prior art* provisions. To prevent the criminal

267

underworld from being handed the details of exactly how to steal money out of an ATM, De La Rue and their client, Barclays Bank, decided to keep the original operating systems secret by not filing for patent. This explains why litigation brought by Smiths Industries, Chubb and others for copyright infringement were dropped in early 1967, just before Barclays Bank unveiled the world's first fully functional 24/7 cash dispenser. In other words, the litigants backed down and acknowledged they were wrong.

Individuals named in each respective patent can be forgiving for believing that it was them that 'invented' the ATM, or aspects of it. But they will have been unaware – until now – that a claim by De La Rue to prior art preceded them, and that Dad, with the support of his team at De La Rue Instruments, can therefore lay legitimate claim to having invented the ATM.

Either way, and whatever your interpretation of history, it must be a rare event when an entire industry can look at a person and say with confidence, as the American company executive said to me that night when listening to my Dad speak, "he started it."

Dad died still unsurprised – and even a little amused – by the industry-wide debate generated by his accolade, as he knew the concept challenged the precepts of 'patent protection,' and thus how the ATM was to be defined.

In every sense, the story resembles that of Edison and the lightbulb: History muddles through to eventually bestow the credit on a single individual even though we know the truth is somewhat different, and that, while the process is as much to do with serendipity as it is to do with inspiration and perspiration, the actual result can only come from the efforts of a team. This, as my Dad was the first to admit, was the case with the ATM. It simply couldn't have been done without the engineers of De La Rue and the bankers of Barclays.

This has been their story as much as it has been my Dad's.

John S-B on starting at De La Rue
c.1950

John S-B, last son of the Raj, in 1931 aged 6, at Cliftonville, Karachi with his
mother, Dorothy Shepherd-Barron, who went on to win the Ladies Tennis
Doubles at Wimbledon three months later.

John S-B doing it the old fashioned way (c.1978)

John S-B, aged 3, on his donkey in Karachi, India (1928) at around the time he met 'Aircraftsman Shaw' and saw a Gold Sovereign for the first time.

Captain John S-B Rawalpindi c.1946

Reg Varney, star of TV show *On The Buses,* making the world's first ATM cash withdrawal at Barclays Bank, Enfield, London, 27 June 1967

De La Rue DACS twin-drawer cash dispenser, June 1967

The Chubb rival to De La Rue's DACS c.1969

NatWest-Chubb reusable Bankcard c.1969

Barclays DACS one-time use Carbon-14-impregnated voucher, May 1968

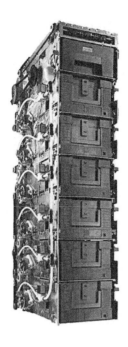

Stack of 6 cash cassettes + dispensing unit

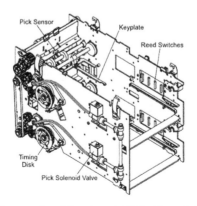

Cross-sectional drawing of ATM pick module

World's lowest ATM
Nottingham, England

World's Tallest ATM
Ankara, Turkey

Finger vein biometric device

Palm vein Biometric device

Iris Recognition Biometric device

World's highest ATM, India

Water ATM, Narobi

World's First Mobile ATM, England c.1970

World's strongest ATM after Hurricane Katrina 2005

World's most edible ATM, Vienna, Austria

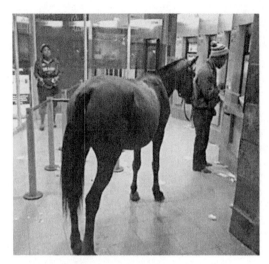

World's most agricultural ATM, South Africa

Security Express Armoured Van (model) c.1968

'Smart' Cash Cassette

'New' denomination US $100 bill, 2015

Euro 50 banknote under UV light

Security Features on New Zealand banknote, 2013

ATM factoids

- Roughly 8.6 billion cash withdrawals are made every month around the world. This equates to around 100 billion transactions every year ... a volume which is forecast to reach 128 billion by the year 2020.

- ATMs dispense a total of $14.1 trillion in cash every year. This equates to almost half-a-million dollars ($450,000) every second.

- The average European ATM serves 1,220 different customers, whose average withdrawal is €122.

- In Europe, an average of €2,562,000 is dispensed per machine per year.

- In the US, a bank ATM averages more than 7,000 transactions per month. This is slightly higher than in Europe where an average ATM conducts 197 transactions every 24 hrs.

- The average consumer uses an ATM 75 times per year i.e a little more than once per week.

- There are over 3.2 million ATMs worldwide, with another being added every three minutes.

- The record for the number of ATM transactions per month is held by Barclays' branch in Camden Town, London with 26,528 made in April 1998, of which 185,000 were made over the three-day Easter weekend.

- The record for the amount of cash disbursed in one month is held by a Barclays ATM located at Terminal 3, Heathrow

Airport in London at £1,605,890 (nearly $2 million) in April 1998.

- $1 million in $100 dollar bills weighs 20.5 lbs or just under 10 kgs.
- There is an estimated $5 trillion in currency circulating the globe.
- A typical wall-mounted ATM houses 18 different modules comprising a total of approximately 11,000 working parts.
- A modern ATM uses more computing power than was available to NASA at the time of the first Moon landings.
- There are 360 billion banknotes in circulation, with 150 billion new ones printed each year. The US prints 38 million dollar bills every day.

awesome ATMs

1. **Strongest:** New Orleans (Hurricane Katrina)
2. **Greenest:** Solar powered, Colombo, Sri Lanka
3. **Cleanest:** Tokyo (steam cleans and irons notes)
4. **Dirtiest:** Brooklyn, New York (most covered in graffiti)
5. **Cleverest:** GCHQ, England (users have to solve an equation within 30 secs before access is allowed)
6. **Smallest:** 42nd St Bagel stall, New York City
7. **Healthiest:** Piccadilly Circus, London (fewest faults)
8. **Thinnest:** Orissa, India (lightest functionality)
9. **Fittest:** Ski-thru ATM, Whistler Mountain, Canada
10. **Tastiest:** Cupcake ATM (Sprinkles, New York)
11. **Highest:** Pakistan / Tibet border
12. **Lowest:** English village
13. **Tallest:** Split, Croatia
14. **Shortest:** Nottingham, UK
15. **Wildest:** South Africa (with horse tethered to ATM)
16. **Holiest:** Vatican City (with instructions in Latin)
17. **Busiest:** Terminal 3, Heathrow Airport (greatest footfall)
18. **Smartest:** Moscow (part Cash Machine, part Robocop)
19. **Hottest:** Basra, Iraq
20. **Coldest:** South Pole Survey Station, Antarctica
21. **Cheapest:** Ankara, Turkey (dispenses coins)
22. **Poorest:** Dadaab refugee camp (Kenya)
23. **Oldest:** London (Barclays, Science Museum)
24. **Most mobile:** Papua New Guinea (containerised)

25. **Most self-sufficient:** Philippines (satellite-linked, solar-powered)
26. **Most edible:** Vienna
27. **Most architectural:** Nepal (in a Stupa)
28. **Most remote:** Solomon Islands
29. **Most state-of-the-art:** New York (latest technology)
30. **Most multi-lingual:** London (uses Cockney rhyming slang)
31. **Most tactile:** almost anywhere (instructions in Braille)
32. **Most colourful:** Bangkok (bright pink)
33. **Most helpful:** Cambodia (foiled a kidnap gang)
34. **Most secure:** Tehran
35. **Most robbed:** Moscow
36. **Least submersible:** Sindh, Pakistan (on pontoons)
37. **Most submersible:** US (nuclear) submarine
38. **Most portable:** On a donkey in Baluchistan
39. **Most complicated (to operate):** Sao Paolo, Brazil
40. **Most northerly:** Trømso in Norway
41. **Most dangerous:** Mogadishu, Somalia
42. **Most Royal:** Buckingham Palace
43. **Most musical:** Louisiana (in a church organ)
44. **Most expensive:** Dubai (gold bullion)
45. **Most personal:** Alice Cooper's car
46. **Most biometric:** Amman, Jordan
47. **Most political:** Manchester, UK (correctly forecast the UK's 'Brexit' vote in June 2016)
48. **Most alternative:** Bitcoin (New York)
49. **Most airworthy:** Pakistan (underslung from a UN helicopter during flood relief operations)
50. **Most secret:** National Security Agency, Washington DC

51. **Most functional:** Number of functions (South Africa >200 inc funeral plan)
52. **Most explosive:** Switzerland (blown up in testing)
53. **Most animalistic:** Housed in a plastic panda, Bangkok zoo
54. **Most crowded:** Turkey (most in one location)
55. **Most entrepreneurial:** On a motorcyclist's back
56. **Most generous:** Paris (spat out wrong notes)
57. **First:** Barclays Bank, Enfield

For photo's and further information on these awesome and 'extreme' ATMs, see: lovecashmachines.org

index

290

Printed in Great Britain
by Amazon